Handbook on Concrete Block Paving

Wasantha Mampearachchi

Handbook on Concrete Block Paving

Springer

Wasantha Mampearachchi
Department of Civil Engineering
University of Moratuwa
Moratuwa, Sri Lanka

ISBN 978-981-13-8416-5 ISBN 978-981-13-8417-2 (eBook)
https://doi.org/10.1007/978-981-13-8417-2

This Springer imprint is published by the registered company Springer Nature Singapore Pte Ltd.
The registered company address is: 152 Beach Road, #21-01/04 Gateway East, Singapore 189721, Singapore

Preface

This textbook is an essential guide to all facets of concrete block paving technology and aims to assist practising engineers, technologists and postgraduate students. Among all the materials required, mechanistic and empirical models are available for design pavements, standards and specifications are available to determine the material and construction process, and engineers and technologists have fairly good knowledge about the traditional pavement types. However, a comprehensive compilation of such technical information on this subject not being available had led those interested in the field to resort to the means of referring to multitudinous resources. Considering those circumstances, this book was composed to reflect the very best of concrete block paving technology thus is inclusive of the proper pavement design standards, guidelines for block selection, mixture design, pavement design and appropriate construction and maintenance procedures that should be undertaken. Simply stated, written by a renounced researcher in the area of concrete paving technology, it covers all aspects of concrete block paving technology from foundation design through to finishing.

Moratuwa, Sri Lanka | Wasantha Kumara Mampearachchi

Acknowledgements

I would like to extend my sincere thanks to all those who supported me through the process of preparing this book. As I believe, without mentioning those who contributed to its success, this book cannot be called ‘complete’.

I grasp the earliest opportunity to acknowledge the commitment of Chamod Hettiarachchi, my former student, in preparing and penning the content under the third chapter of the book entitled ‘Construction Process’.

On the same note, I must say I am deeply indebted to Prof. Mangriker Gunaratne, the Chairperson, Department of Civil and Environmental Engineering, University of South Florida, who reviewed the entire book. His comments were extremely helpful as they paved the way to improve the technical content, organization and readability of the text. Furthermore, I thank Prof. Saman Bandara, Professor of Civil Engineering, Department of Civil Engineering, University of Moratuwa, who edited the final manuscript.

Moreover, though the materials included in this handbook were prepared based on the research studies I conducted myself, it would be my greatest pleasure to acknowledge the former master's students, W. P. H. Gunaratne, A. A. Senadeera and Aruna Karunaratnne, and the former undergraduate research students, U. S. Pilanavithana, P. S. Perera and R. S Appuhami, who have immensely contributed to the developments in the concrete block paving research at the University of Moratuwa.

Also, I appreciate the assistance given by Akila Hettiarachchi and Samitha Amerasinghe, Technical Assistants who carried out the editing and typing. Dhanushika Gunathilake and Chamod Hettiarachchi, Graduate Research Assistants, Department of Civil Engineering, University of Moratuwa, who edited the final draft and provided figures and corrections as required, have also been a great support. And I wish to thank Sanka Amerasena and the students of Highway Engineering Society Student Chapter of the University of Moratuwa who provided me with the assistance I needed by preparing all the figures. Furthermore, I would like to thank Eshara Randeniya for supporting me with the proofreading and Uditha Galgamuwa for designing the cover page.

Last but not least, I would like to thank my family for having my back throughout every step I take. Without the help of all, this book would not have been a possibility, so I once again express my humble gratitude to all.

Contents

List of Figures

List of Tables

Chapter 1
Introduction

1.1 Why We Need Concrete Block Paving

In most countries, a high percentage of roads have been constructed mainly using asphalt and concrete. However, people, realizing defects in these roads gave rise to the idea of having other options for constructing roads. Since the unit cost of a concrete road is very high, it is quite unbearable for countries with a developing economy.

There are several disadvantages to asphalt pavement roads. Bitumen; the primary material of asphalt, comes as a byproduct of purified crude oil. Therefore the prices of bitumen also increase parallel to the escalating crude oil prices. The most vital issue is that bitumen may not be a substantial construction material any more, due to the rapid depletion of crude oil. As a result of that, the road designers try to improve the existing road construction technology to suit the present era. This 'Concrete Block Paving' (CBP) concept is based on the concept of 'Stone Paving' which lasted in the world thousand years back.

There are only few concrete block pavements in Sri Lanka. The Road Development Authority and Provincial Road Authorities are going to promote CBP for rural roads as an alternative for bitumen roads, considering the benefits such as low construction and maintenance cost, aesthetic appearance, and durability.

1.2 Historical Background

The history of stone paving roads dates back to 4000 BC, Assyria. By 2000 BC flagstones were being used to pave village streets (Concrete Manufacturing Association, 2004). Cobble stones which were water worn stones or large pebbles about 150 mm had been used for paving at a very early stage and with the passage of the time were replaced by hand cut stones. The most famous of all Roman roads is the Appian Way, built by Roman engineers in 312 BC. It still carries traffic

W. Mampearachchi, *Handbook on Concrete Block Paving*,
https://doi.org/10.1007/978-981-13-8417-2_1

Fig. 1.1 The early interlocking pavement—Roman Appian way (ICPI, 2004)

between Rome and Italy's south-eastern port of Brindisi. This was surfaced with tight-fitting paving stones (Fig. 1.1).

Later on, people started to use burnt bricks for road construction as an alternative to stone paving. The modern version of concrete paving blocks (pavers) were manufactured in Netherlands in the late 1940s as a replacement for burnt bricks, because there was a scarcity of bricks due to the Second World War. Concrete blocks were introduced as an alternative to brick paving and subsequently became recognized as a durable paving material.

In the very early stages, the block paving concept was put into practice due to the aesthetic appearance. However, with time, designers identified the significant advantages of concrete block paving over other techniques. As a result, concrete block paving has been extensively used for commercial, municipal and industrial applications since the 1960s. The study of Ghafoori and Mathis (1998) reveals that over a hundred million square meters are paved annually in Europe, while it is estimated that more than two hundred eighty million square meters of concrete paving units are produced worldwide (Ghafoori & Mathis, 1997).

1.3 Applications of the Concrete Block Paving

Applications of CBP can be categorized based on the traffic condition;

Non-Traffic: Building premises, Foot paths, Shopping Malls, Pedestrian walks, Landscapes, Public gardens, Domestic paving, Embankment slopes etc.

Light Traffic: Car parks, Office driveways, Commercial complexes, Rural roads, Residential streets etc.

Medium Traffic: City streets, Small market roads, Utility service stations.

Heavy and Very Heavy Traffic: Container/Bus terminals, Ports/Doc yards, Industrial complexes, Heavy duty roads, Bulk cargo handling areas, Factory floors/Pavements, Airport pavements etc.

Non-Traffic

Domestic use: Concrete Block Paving is used in domestic applications. It provides a very attractive appearance and enhances the value of the property (Fig. 1.2).

Pedestrian walkways: Concrete Block Paving can be used in pedestrian walkways to provide an aesthetically pleasing appearance (Fig. 1.3).

Embankment protection: The use of CBP alongside the freeways is a very effective and quick method of slope protection (Fig. 1.4).

Fig. 1.2 CBP in Domestic use in a temple, Beruwala and Mihindu Seth Medura, Attidiya, Sri Lanka

Fig. 1.3 Pedestrian Walkway at Waters Edge, Sri Lanka

Fig. 1.4 Embankment protection at Colombo Katunayake Expressway, Sri Lanka

- **Light Traffic**

Residential streets: The flexibility of the block pavement in adapting to changes in land use and road geometry, coupled with their highly aesthetic nature and low maintenance costs, has firmly established block paving as a viable solution for residential street construction (Fig. 1.5).

Rural roads: The present Sri Lankan government and the Road Development Authority has focused on introducing concrete block pavements for rural roads as a cost-effective paving method (Figs. 1.6 and 1.7).

Car parks and driveways

See Fig. 1.8.

- **Medium Traffic**

City streets: The ease of access to underground services, the ability to withstand heavy slow moving traffic, the low maintenance costs and the aesthetic advantages of

Fig. 1.5 Residential street at Lund City, Sweden

Fig. 1.6 A rural road in Mahawa, Sri Lanka

Fig. 1.7 Borupana- Ratmalana Road- Sri lanka

interlocking concrete pavements have established block paving as a major alternative for the surfacing of city streets (Fig. 1.9).

- **Heavy and Very Heavy Traffic**

Industrial hard surfacing: The most cost-effective use of block paving can be found more frequently in the industrial sphere. The ability of the block pavement to withstand concentrated heavy load and to resist the wheel load of heavy off-road vehicles such as forklifts, straddle carriers and trans trainers, have made interlocking concrete paving the first choice for many industrial pavements, including in particular, container yards (Figs 1.10 and 1.11).

Bus terminals: Typically buses stop and start continuously at the same locations, therefore, can quickly cause rutting in flexible pavements. This problem is aggravated by the spillage of oil and lubricants typical of bus operations, which leads to the fluxing and softening of asphaltic materials. To overcome these problems, interlocking concrete block paving is increasingly used.

Fig. 1.8 BMICH car park and Cinnamon bay driveway, Beruwala, Sri Lanka

Fig. 1.9 City Street in Malmo City—Sweden

Fig. 1.10 A Container Yard (Concrete Manufacturing Association 2004, bk. CBP 1 Introduction)

Fig. 1.11 A Warehouse (Concrete Manufacturing Association 2004, bk. CBP 1 Introduction)

Often such facilities are located on reclaimed lands, subjected to ongoing settlements. The ability to recycle (replace) block pavement facilitates the use of concrete block paving in such places because settlement deformation can be easily remedied (Fig. 1.12).

1.4 Advantages of Concrete Block Paving

There are many advantages of concrete block paving over other paving methods. They are,

- Simple construction
- Provision for underground repairs

Fig. 1.12 Bus Terminal - Lund City, Sweden

- Long lasting surface
- Reusable
- No adverse thermal effect
- The possibility of allowing traffic just after construction or repair
- High-quality control
- Higher elastic deflection
- High skid resistance
- Low maintenance.

1.5 Limitations

Concrete block paving can be expensive if specialist block materials are chosen, good practices are followed in the preparation of subbase/base and laying, and the edge restraints are provided. There are only a few limitations in concrete block paving. They are,

- Not suitable for streets where the speed limit is above 70 km/h (due to high noise and vibration)
- Need edge restraints or curb lines
- Manual laying consumes more time.

Chapter 2
Structural Behaviour of Concrete Block Paving

2.1 Interlocking Mechanism

When the tire load is applied on top of the concrete block pavement, concrete blocks (pavers) tend to rotate and translate. Block rotation and translation are generally associated with hinge formation as shown in Fig. 2.1.

'Interlock' is the inability of pavers to move independently from its neighbours. When considering the design and construction of block pavements, three types of interlocking must take place simultaneously: such as vertical, rotational and horizontal interlocking as illustrated in Fig. 2.2.

Vertical interlock is achieved by the shear transfer of loads to surrounding units through the sand in the joints. Rotational interlock is maintained by pavers of sufficient thickness, placed closely together, and restrained by a curb from the lateral forces of vehicle tires. When progressive stiffening has been stabilized, the pavement experiences what is known as a "lockup." Horizontal interlock is primarily achieved through the use of laying patterns that disperse forces from the braking, turning, and accelerating of vehicles. They maintain the horizontal interlock, while the units are

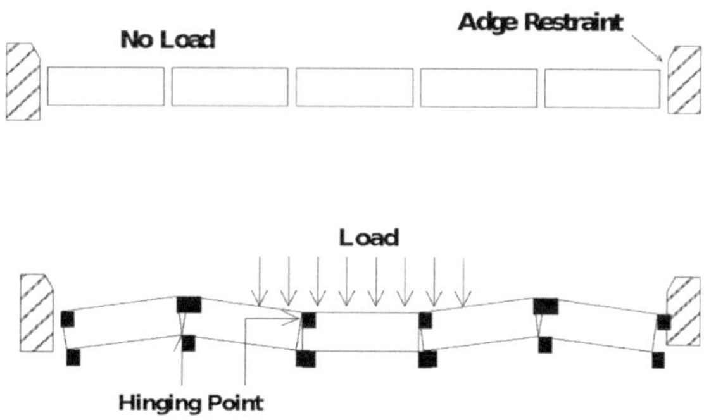

Fig. 2.1 Deflection shape of block pavement (Panda & Ghosh, 2002)

W. Mampearachchi, *Handbook on Concrete Block Paving*,
https://doi.org/10.1007/978-981-13-8417-2_2

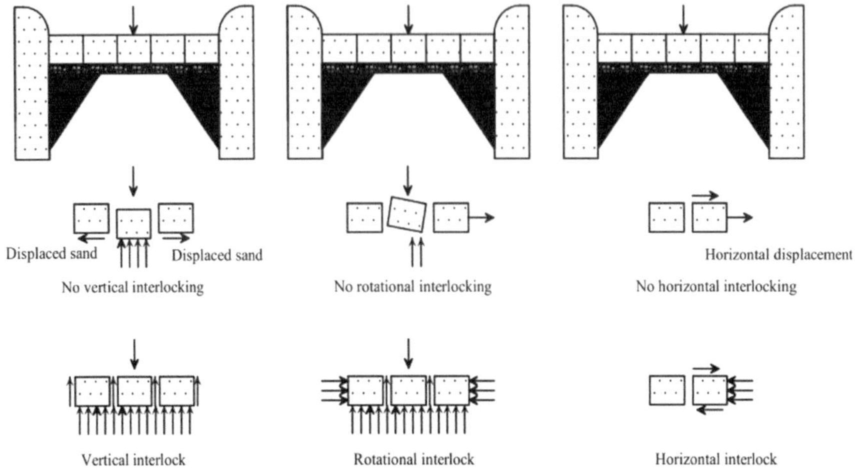

Fig. 2.2 Type of Interlocking (ICPI, 2004)

subjected to repeated lateral loads from vehicle tires. Stable edge restraints such as curbs, are essential for the whole system to be locked up.

2.2 Shape of the Block

Even though around 40 block shapes have been patented, all of these are not suitable for traffic pavements. The common block types used in the road construction industry are shown in Fig. 2.3. They are; (1) Uni Style, (2) Key Stone, (3) Cobble, (4) Clover, (5) Honey Style and (6) Satin Pave.

Based on past experience and researches, the block paving industry has been involved in producing a vast range of block shapes. Concrete blocks can be divided into three categories based on the shape (Concrete Manufacturing Association, 2004). Extent of Interlocking can vary with the shape of the block. They are named as type 1, type 2 and type 3,

Type 1 blocks allows a geometrical interlock between all vertical or side faces of adjacent blocks (e.g.—Uni style).
Type 2 blocks allows a geometrical interlock between some faces of the adjacent block (e.g.—Key stone).
Type 3 blocks allows no geometrical interlock between adjacent blocks (e.g.—Cobble type).

Research have shown that 'type 1 blocks' develop the best resistance to both the vertical and horizontal creep, and therefore is generally recommended for all industrial and heavy duty applications. Type 2 and 3 blocks are generally selected for aesthetic reasons.

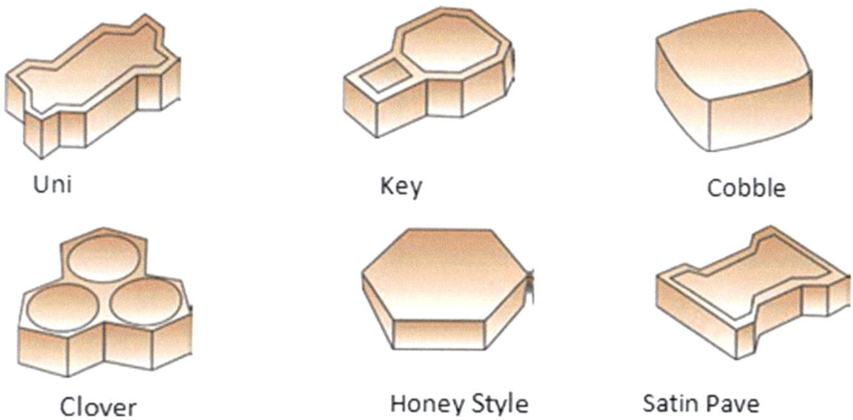

Fig. 2.3 Common block types used in CBP

Blocks with a complex shape have a larger vertical surface area than rectangular or square blocks of the same plan area. Consequently, rectangular or square shaped blocks have larger frictional areas to allow load transfer to adjacent blocks.

2.2.1 *Field Performance of Block Shapes*

According to the traffic studies done in Australia, South Africa and USA, block pavements laid with shaped (denoted) blocks have been reported to perform better than pavements laid with rectangular units. Generally, pavements laid with denoted pavers tend to exhibit smaller deformations than pavements utilizing rectangular pavers. Recent Australian studies show that paver shape also influences the development of the horizontal creep under traffic. Tests show that for a given laying pattern, rectangular pavers are associated with a much greater horizontal creep than denoted units.

A field test was conducted in the test road at Mahawa, Kurunegala, Sri Lanka to investigate the effect of block shapes, laying patterns and laying angles, on the performance of CBP. The deflection information from the Benkelman beam test, gathered for different block shapes are given in Fig. 2.4. The average deflection in all three types of concrete blocks within the test area, for a standard axle of 8160 kg, were 49, 54 and 54 for Uni style, Cobble and Key stone, respectively. Here, the laying angle was kept at 90° and the stretcher block pattern was used. The block shapes were changed to evaluate the deflection. It was observed that Uni style blocks show the least deflection while the Cobble shape blocks and the Key stone blocks have the most deflection.

Furthermore, the Benkelman beam deflection data were collected at different locations from the cross section, to plot the deflection basin for an applied load.

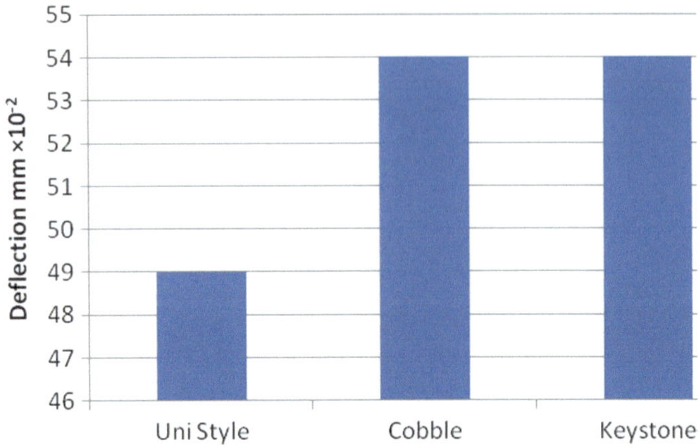

Fig. 2.4 Effect of block shapes on deflection

These data were collected in a transverse direction of the road, by loading the road at the centerline and measuring the deflection at 0 mm (at the center), 300, 600 and 900 mm from the center. Cross sectional deflection profiles (deflection basin) are shown in Fig. 2.5. A narrow deep deflection basin was observed for the cobble block shape, with the highest deflection at the center, and a deflection close to zero at 900 mm from the center. The deflection of the Uni style block shape was lesser at the center of the road, but extended a longer distance than the Cobble and Keystone block shapes. A Uni style block is a type 1 block, and has a geometrical interlock between adjacent blocks in all the vertical faces. Consequently, Uni style blocks have larger frictional areas for load transfer to adjacent blocks. It is postulated that the effectiveness of the load transfer depends on the vertical surface area of the blocks.

The gaps between the concrete blocks vary when vehicles move on the road. A field testing was conducted to measure the gaps between blocks at predetermined locations for selected block shapes. The gap variations of different block shapes, with time (number of vehicle passes over the blocks) are shown in Fig. 2.6. The joint gap variations of three different block shapes which are laid along the road do not show any significant variation until 7.9E + 03 ESAL. However, the Uni style block shape has a lesser gap variation after 7.9E + 03 ESAL, and this gap variation becomes constant after 15.9E + 03 ESAL. The Cobble block and the Key stone block have larger gap variations, but the variations are of the same magnitude.

The Cobble block shape has the poorest performance while the Uni style block shape has the best performance with respect to the deformation (lead to permanent deformation in wheel paths, rutting) and joint gap variation. Uni style blocks have a geometrical interlock between all side faces of adjacent blocks causing the minimum joint gap variation. Therefore, it is possible to conclude that the Uni style block shape has the best performance when used as a block pavement under both horizontal (braking action) and vertical loading.

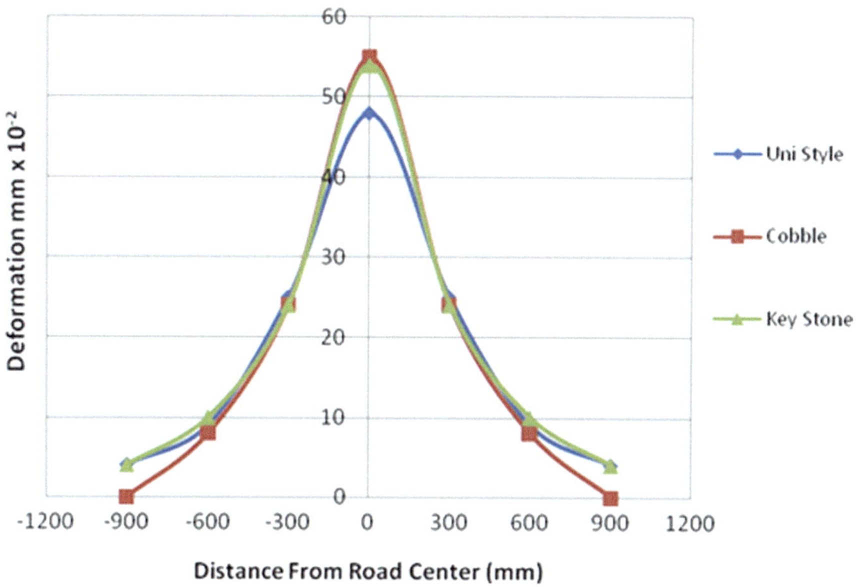

Fig. 2.5 Effect of block shapes on deflection profile

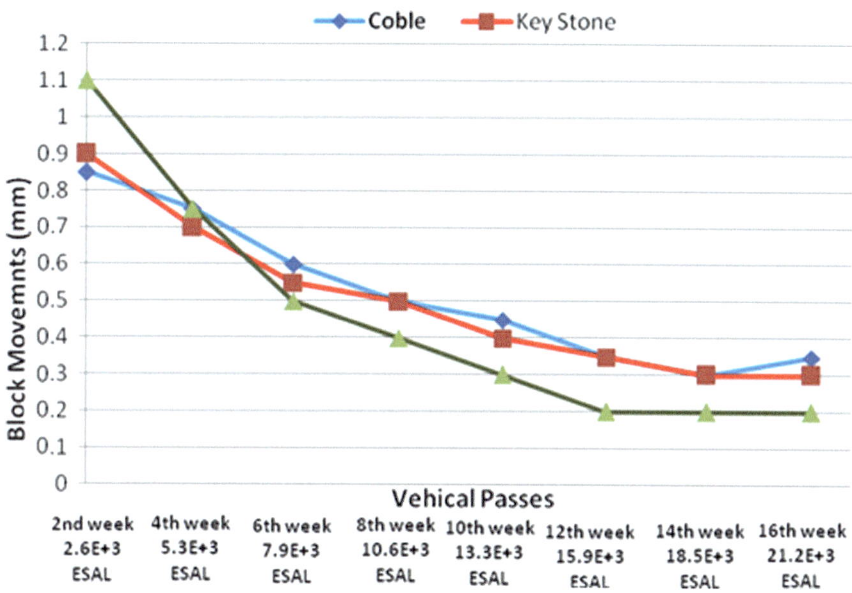

Fig. 2.6 Joint gap variation with time

2.3 Thickness of the Block

Some of the early studies on concrete block paving, done by (Clark, 1978) reported a small improvement in the pavement performance with the increase of block thickness. Miura, Takaura, and Tsuda (1984), Shackel (1980), and Shackel, O'Keeffe, and O'Keeffe (1993) claimed that an increase in the block thickness reduced elastic deflection, and the stress was transmitted to the sub base. The recent research conducted by Panda and Ghosh showed a significant reduction in the elastic deflection as the block thickness increased. Rectangular blocks of the same plan dimension, with three different thicknesses: 100, 80 and 60 mm, were tested in a prototype pavement model. Blocks were laid in a stretcher bond pattern in the prototype model and the deflection related to the applied load was measured. Load was varied from 0 to 50 kN. It was observed that the shapes of load—deflection curves were similar for all the selected block thicknesses. A change in thickness from 60 to 100 mm, significantly reduced the elastic deflection of the pavement. The comparison is shown in Fig. 2.7.

Thicker blocks provide a higher frictional area. Thus, the load transfer will be high with thicker blocks. The individual block translation in thicker blocks is more with the same amount of block rotation. As a result, the back thrust from the edge restraint will be high. The thrusting action between adjacent blocks at hinging points (Panda & Ghosh, 2001) is more effective with thicker blocks. Thus, deflections are lesser for thicker blocks. The combined effect of the higher friction area and the higher thrusting action in thicker blocks provide a more efficient load transfer. Thus, there is a significant change in the deflection values, when increasing the thickness of blocks. It can be concluded that the response of the pavement is highly influenced by the block thickness.

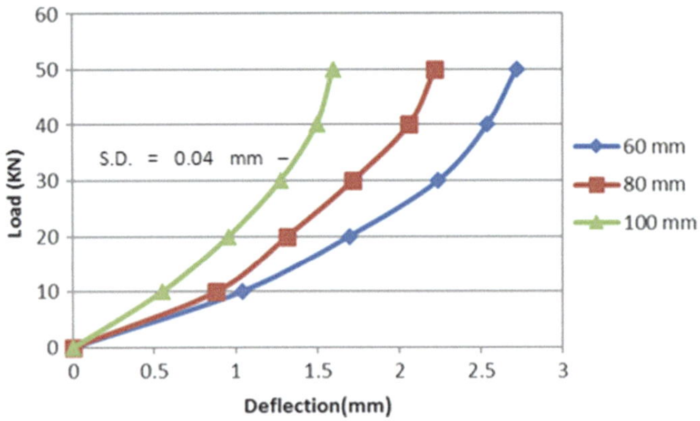

Fig. 2.7 Effect of block thickness on behavior of block pavement

2.4 Block Size

Shackel (1980), reported that the load-associated performance of block pavements was essentially independent of the size of blocks, but Panda and Ghosh (2002) proved that there was a clear response on the concrete block pavement performance with the variation of block sizes. For this experiment, Panda and Ghosh (2002) used rectangular blocks laid in the stretcher bond pattern. The responses were measured for different thicknesses under the applied loads. The shapes of the load deflection paths are similar for all block sizes. Lesser deflections of the pavement have been obtained with the increase of the block size (Fig. 2.8). The number of joints per unit area of pavement is high when small blocks are used. The translation of a block in the loading area cannot be effectively reflected to its full extent at the edge restraints. This leads to the lesser compression of jointing sand, and lowers the buildup of joint stresses. Thus, the pavement experiences a higher deflection under the same load when smaller blocks are used. It is established that block size influences the load-spreading ability of the pavement. The larger the size of the block, the more the pavement performance will improve.

2.5 Block Strength

In 1980, Shackel reported that the load-associated performance of block pavements was essentially independent of the compressive strength of the blocks, and that was verified again by Panda and Ghosh (2002).

Panda and Ghosh (2002) conducted the experiment using three types of blocks with identical sizes, made out of different concrete mixtures, to have different compressive strengths. The areas of friction were the same for the three block types. Each individual concrete block was subjected to compressive stress with a negligible bending stress because their sizes were small. The elastic modulus of the concrete block

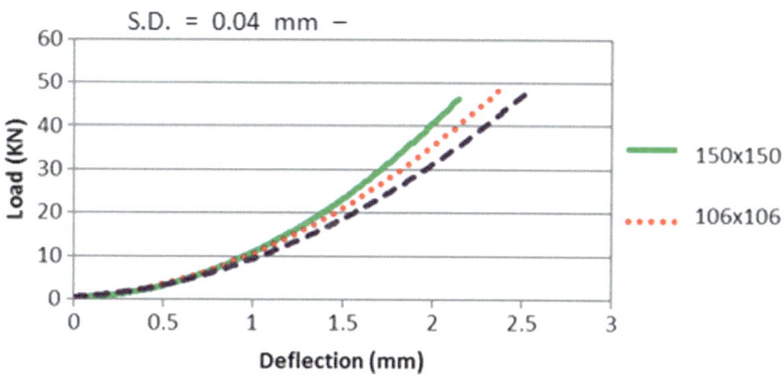

Fig. 2.8 Effect of block sizes on behavior of block pavement

was many times higher than that of the underlying materials. The concrete blocks behaved as rigid bodies in the pavement, and transferred the external load to the adjacent blocks and underlying layers by the virtue of its geometrical characteristics, rather than its strength. Panda and Ghosh (2002) established that the load-associated performance of block pavements was independent of the compressive strength of blocks.

Selection of materials and suitable mix proportions are discussed in Sect. 3.1.

2.6 Laying Pattern

Load spread in concrete block paving roads is accomplished by the interlocking actions. Hence, the laying pattern of pavers (concrete blocks) act as a major component in the load spreading in concrete block paved roads. Therefore, it is very important to identify the most effective laying pattern which can improve the interlocking action.

Different laying patterns are commonly used for block laying work. Knapton (1976) discovered that the laying pattern did not significantly affect the static load-spreading capacity of the pavement. From plate load studies, Miura (1984) and Shackel (1993) have reported that, blocks laid in a herringbone bond exhibited a higher performance than the stretcher bond for a given shape and thickness. In 2001 Panda and Ghosh reported that no significant variation can be found in the performance of block pavements with the differentiation of laying patterns. Therefore, it is apparent that the findings are inconsistent about the performance of block pavements with respect to different laying patterns. Therefore, an exploration regarding this factor in concrete block paving is essential.

There are mainly four types of block laying patterns used in industrial applications: such as the herringbone bond, stretcher bond, basket weave and stack bond. In the concrete block laying practice, 45° angle patterns are used. However, pavers perform better when their laying pattern runs in line or at a right angle to the direction of traffic. When vehicles deviate by even an angle of 10° on any direction, interlocking action becomes less effective (Murat Algin, 2007). Hence, it is recommended not to use 45° angle laying patterns for concrete block laying work in road construction. Concrete block laying patterns commonly used in road industry are shown in Fig. 2.9.

2.7 Determination of an Effective Laying Pattern

According to the 3-D finite element software model of Mampearachchi and Gunarathna (2010), the herringbone bond pattern gave a lower vertical deflection against loading, compared to other bonding patterns. Therefore, it is apparent that the herringbone bond is more stable against vertical loading (Fig. 2.10). Figure 2.15 clearly shows that herringbone bond pattern gives minimum

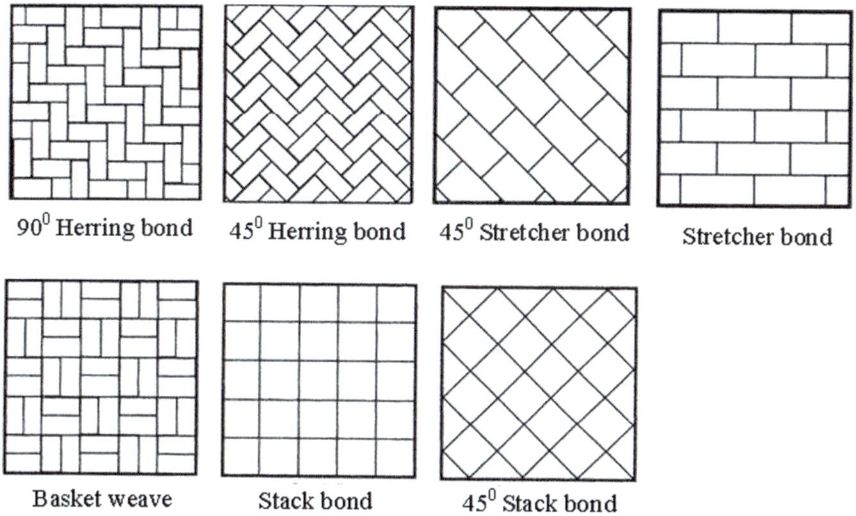

Fig. 2.9 Concrete block laying patterns

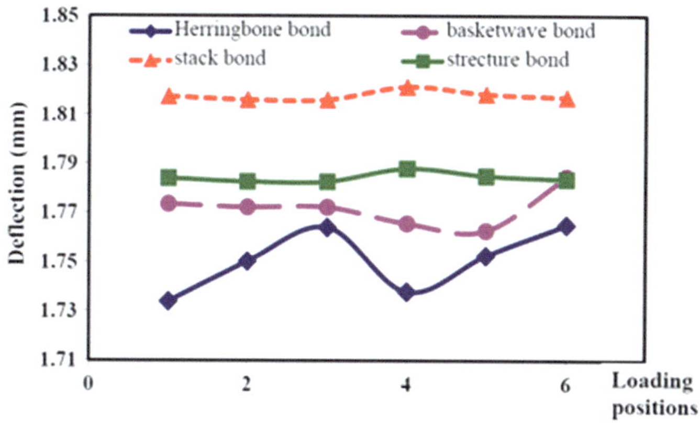

Fig. 2.10 Loading position and vertical deflection in different laying patterns (effect of vertical load) (Mampearachchi & Gunarathna, 2010)

vertical deflection against the breaking action of vehicles compared to other laying patterns. However, the stretcher bond pattern gives minimum horizontal deflection against breaking action, compared to other laying patterns (Fig. 2.12). The horizontal deflection values were less than 0.7 mm and vertical deflection (Fig. 2.10) was a minimum of 1.82 mm. So, vertical deflection is nearly three times higher than horizontal deflection for breaking action. Therefore, the vertical deflection curve can be considered as the deciding factor for selecting the stronger bond pattern under the breaking effect in that study (Fig. 2.11).

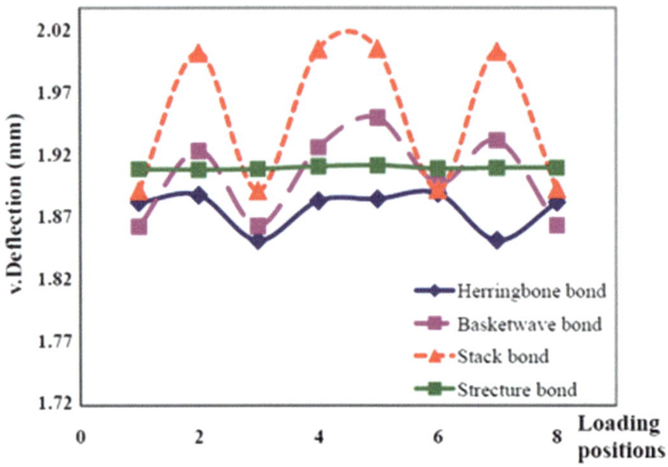

Fig. 2.11 Loading positions versus vertical deflection in different laying patterns (effect of breaking action) (Mampearachchi & Gunarathna, 2010)

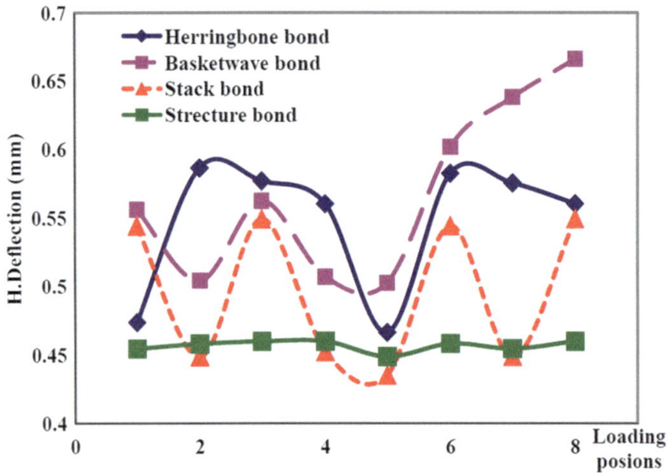

Fig. 2.12 Loading position versus horizontal deflection indifferent laying patterns (effect of breaking action) (Mampearachchi & Gunaratna, 2010)

Tests for the field performance of the selected block shapes were conducted by (Mampearachchi & Senadeera, 2014) to verify the FEM finding, since there is a contradictory behavior of block patterns for vertical and horizontal deflection against breaking action.

2.8 Field Performance of Laying Pattern

Laying Pattern used:

A total of 100 m road length was selected and divided into 10 equal parts, and blocks were paved according to selected combinations in each different section, as given in Fig. 2.13.

There are four different block laying patterns employed in the test section for the comparison of deflections. Herring, stretcher, basket and stack laying patterns were compared based on the data collected from test sections. In this case, the laying angle remained constantly at 90° and the block shape used for the test was Uni style. Then the laying pattern was changed, and the deflections were measured at the center of loading and locations were selected to capture the shape of the deflection basin.

Figure 2.14 shows that the Stack bond has the largest deflection. The Stretcher bond type and the Herringbone bond type have an equal amount of deflection, which is also the lowest deflection. Therefore, these two types of patterns are more suitable for a road.

The data collected for the transverse direction of the road, shows how deflection is distributed within the road section. Figure 2.15 shows that the Stretcher bond type and the Herringbone bond type have the least deflection. Also the variations of deflection among different types of laying patterns, are higher than the variations of deflection of different block shapes and angles. The largest deflection occurs at the center of the road for the Stack bond. The deflection of the Basket weave is larger than the deflection of Herringbone bond and Stretcher bond, but it is lower than that of the Stack bond. However, the deflection of Stack bond reaches zero much quicker than the other three bond patterns. The deflections of other three patterns continue further than the Stack bond pattern, though they have lesser deflection at the center of the road. The load transfer from block to block is the reason for the spreading of deflection for an extended length from the centerline.

The block joint gap variations among four different paving patterns show how the gaps increase or decrease due to vehicular loading. Figure 2.16 shows that the Stretch and Herringbone bonds are nearly the same, and also have the least joint gap variations. It shows that Stack bond pattern has the least gap variation at the beginning but becomes the highest gap variation after 1.06E + 4 ESAL. The gap variations for all bond types become nearly constant after 1.85E + 4 ESAL. Therefore, it is concluded that Stretch and Herringbone bond patterns show the best performance of the block pavement under load, with respect to all the parameters considered in this study.

The analysis mentioned above can be summarized to evaluate the results, in order to find the best selection, as given in Table 2.1.

Type of block - Uni Style
Pattern - Stretcher bond
Angle - 0^0

Type of block - Uni Style
Pattern - Stretcher bond
Angle - 45^0

Type of block - Uni Style
Pattern - Stretcher bond
Angle - 90^0

Type of block - Uni Style
Pattern - Strack bond
Angle - 90^0

Type of block - Uni Style
Pattern – Basket weave bond
Angle – $0/90^0$

Type of block - Cobble
Pattern - Stretcher bond
Angle - 0^0

Type of block - Cobble
Pattern - Stretcher bond
Angle – 45^0

Type of block - Cobble
Pattern - Stretcher bond
Angle - 90^0

Fig. 2.13 Field test laying patterns

Type of block – Key stone

Pattern - Stretcher bond

Angle - 90^0

Type of block - Uni Style

Pattern - Herringbone bond

Angle – $0/90^0$

Fig. 2.13 (continued)

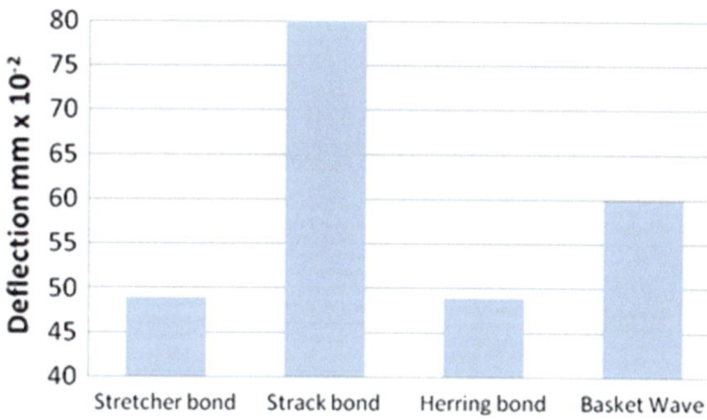

Fig. 2.14 Effect of block laying pattern on deflection

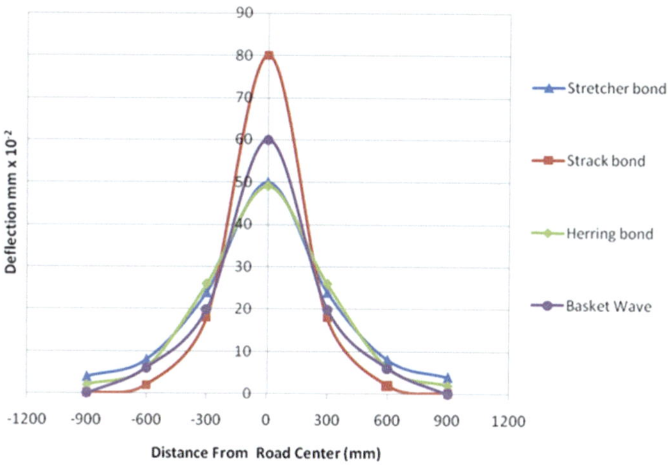

Fig. 2.15 Effect of block laying pattern on deflection profile

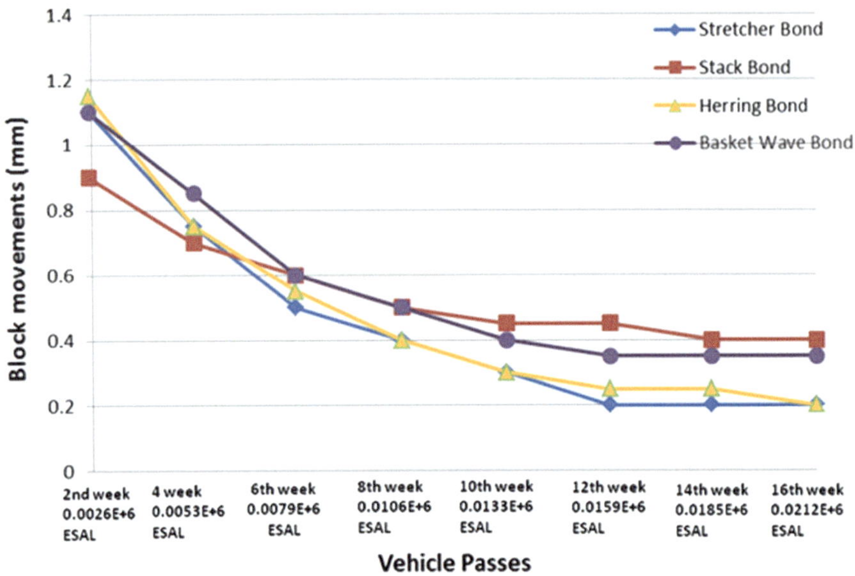

Fig. 2.16 Effect of block laying pattern on block displacement

Table 2.1 Summary of performance of blocks

Variable		Performance			
		Deflection	Deflection profile	Joint gap variation	Selection
Block shape	Uni style	✓	✓	✓	Uni style
	Cobble				
	Keystone				
Laying angle	Angle 0°	✓	✓	✓	Angle 0°
	Angle 45°				
	Angle 90°	✓	✓	✓	Angle 90°
Block laying pattern	Stretcher bond	✓	✓	✓	Stretcher bond
	Stack bond				
	Basket wave				
	Herringbone bond	✓	✓	✓	Herringbone bond

Note Best performance is indicated as ✓

2.9 Joint Between the Blocks, and Filling of Block Joints

CBP is constructed of individual blocks of brick-sized units, placed in patterns with close, un-mortared joints on a thin bed of sand. The joint spaces are then filled with sand. A review of the literature revealed a wide range of views regarding the specifications for bedding and jointing sand, and their contribution to the overall performance of the pavement. Therefore it is important to carry out a comprehensive study about bedding sand and filling sand, and their contribution to the overall performance of the Concrete Block Pavement.

The concrete blocks in the pavement, without jointing sand, behave as individual units. Individual blocks do not transfer the applied load to adjacent blocks. Thus, the block layer has little load spreading capacity. The block layer obtains the load spreading capacity when the individual blocks are interconnected. In 2002 Panda and Gosh verified that pavements without joint sand deflect three times more than pavements with joint sand. Therefore, joints between the blocks should be filled with sand to obtain the full interlocking action in the concrete block pavements.

Jointing sand is the main component of concrete block paving through which load is transferred to the larger area of lower layers by virtue of its shear and dilatancy property (Shackel, 1980). Frictional resistance is developed in the joints under load, preventing the blocks from undergoing excessive relative displacements, and transmitting part of the load to adjacent blocks. The small shear displacements, relative to each other, facilitate the generation of horizontal forces between the blocks, caused by dilatancy in the jointing sand. As a consequence, concrete block paving is capable of achieving substantial distribution of load among neighboring paving units, due to the increased frictional resistance. Crusher-run sand with the same gradation is the best source, rather than river and quarry sand as joint sand for Concrete block paving (Panda & Ghosh, 2001). Because more coarse and more angular sand have a higher dilatancy and a higher angle of shear resistance (Panda & Ghosh, 2002). Nevertheless, in Sri Lanka, river sand has become more popular as a filling material because it is abundantly available.

In concrete block paving, joint width should be compatible with the particle size distribution of filling sand, for improved pavement performance. Panda and Gosh (2002) did laboratory scale experiments to investigate the effect of changing the parameters of bedding sand and joint sand against joint width, on pavement performance.

For this experiment, sand with four gradations conforming to zone I, II, III and IV grading limit Table 4 of Indian Standard 383-1970(RIS1970) were used. Sand samples ranging from coarser to fine in seven different gradations were prepared, using controlled mixes of the main four gradation types, and 'zone I' sand was used in the experiment as bedding sand with a 50 mm loose thickness.

Figure 2.17 shows the response of the pavement for design joint widths, with varying qualities of sand from zone I to zone IV in the joints (Panda and Ghosh, 2002). As the joint width decreases, the deflection of the pavement also decreases for zones III and IV (Type 6 and 7) of jointing sand. With jointing sand in zones I

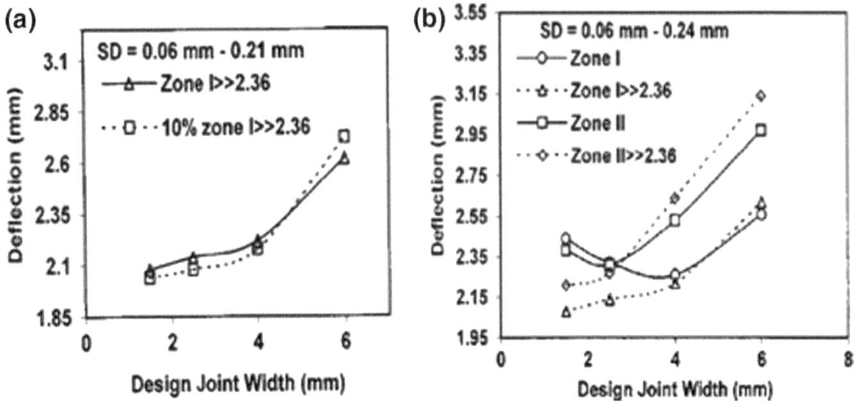

Fig. 2.17 CBP deflection for jointing sand with varying joint width (Panda & Ghosh, 2002a)

and II (Type 1 to 5), the deflection of the pavement decreases up to a certain point and then slightly increases with the increasing joint width. The higher the thickness, the lesser the normal stiffness of the joint will be. This will lead to more rotations and translations of blocks. Thus, there will be more deflections under the same load, for thicker joints.

Some of the grains, coarser than the joint width, were unable to slip inside and that lead to the optimum joint widths in Zone I and II (Type 1 to 5) to be 3 and 4 mm, respectively (Fig. 2.17). A large amount of sand remained outside the joint showing sand heaps on the block surface. The coarse grains of sand choked the top surface of joints and prevented the movement of other fine grains in to the joint. Therefore, there might be loose pockets or honeycombing inside the joint. The joint stiffness decreases, and in turn reflects slightly higher deflections. At the optimum joint width, there is a high possibility that single grains of average size, close to the joint width, will retain in the joints during joint filling. As the average grain size for zone II (Type 4 and 5) sand is less than that of zone I, the magnitude of the optimum joint width is less for zone II (Type 4 and 5) jointing sand.

In general, for constant joint width, the deflection of the pavement increases when the grading of sand is varied from zone I to zone IV (Type 1 to 7). That is, the coarser the sand, the higher the performance (Panda and Ghosh, 2002). This is because of the decrease in shear strength of sand from zone I to zone IV (Type 1 to 7) (Table 2.2) and it is compatible with the findings of Panda and Ghosh (2001) which state that coarser sand provide high shear resistance and higher dilatancy. These findings are similar to those observed by Knapton and O'Grady (1983).

The comparison of deflections of the pavement using Type 2 and Type 3 sand of zone 1 are shown in Fig. 2.17a. Lower deflection is maintained up to a joint width of 4 mm in both Type 2 and Type 3 sand. A comparison of Type 1 and Type 2 sands of zone 1, and type 4 and 5 of zone 2, is shown in Fig. 2.17b. It is clear that zone 1 sand is more effective as jointing sand. Type 2 and Type 3 sand are more effective

Table 2.2 Used sand gradations for research (Panda & Ghosh, 2002)

Sieve size (mm)	Zone I			Zone II		Zone III	Zone IV
	Type 1	Type 2	Type 3	Type 4	Type 5	Type 6	Type 7
	% passing	100% passing 2.36 mm	100% passing 2.36 mm 10% retain 0.075 mm	% passing	% passing 2.36 mm	% passing	% passing
20	–	–	–		–	–	–
10	100	100	100	100	100	100	100
4.75	94	100	100	100	100	100	100
2.36	70	100	100	88	100	98	99
1.18	46	65	69	72	81.81	89	94
0.60	30	42	48	45	51.13	63	88
0.30	16	22	29	21	23.86	25	38
0.15	6	8	16	6	6.81	8	6
0.075	2	2	10	2	2.27	3	3

jointing sand up to a joint width of 4 mm, and Type 1 sand is better to use in joints of more than 4 mm. However, sand more than 15% retained on 0.075 mm sieve has shown fine sand sticking to the surface and preventing the entry of remaining sand.

Gradation limits of Type 1–Type 7 and fine aggregate specification for concrete, ASTM C-33 are shown in Fig. 2.18. Both zone 1 and 2 sand lie within the limits of ASTM C-33 which is recommended for both jointing and bedding sand. From the findings of the research, it is better to select the sand in the lower limit of ASTM C-33 which is closer to zone 1 sand recommended by Panda and Ghosh (2002). However, the selection of gradation of the jointing sand should comply with the joint width.

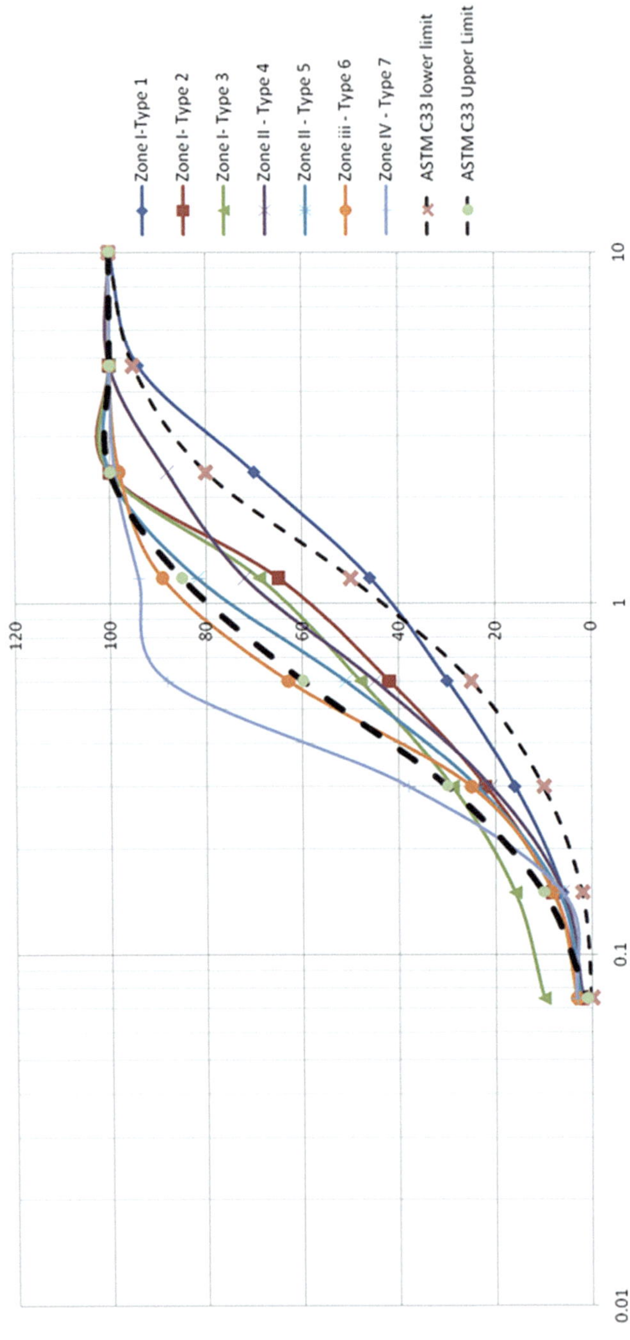

Fig. 2.18 Gradation limits of selected sands

Chapter 3
Construction Process

3.1 Mix Design of ICBP

Interlocking concrete block pavers are used in a wide range of applications covering simple garden patios to heavy-duty container terminals. Hence, the production of blocks in accordance to strength and durability requirements is a major concern. According to standard specifications on block pavers around the world, various strength classes and block thicknesses have been specified for various applications.

According to IS:15658:2006 (Institution, I.S., 2006) and SLS 1425 Part II (Institution, S.L.S., 2011), there are four strength classes recommended based on the application. The recommendations given by IS:15658:2006 on "Precast Concrete Blocks for Paving—Specification" are presented in Table 3.1.

However, ASTM C936 Standard Specification for Solid Interlocking Concrete Paving Units (ASTM International, 2018) requires a minimum average of 8000 psi (55 MPa) with no individual unit below 7200 psi (50 MPa). Concrete paver sizes are defined in this standard as having a minimum thickness of 23/8 in. (60 mm), an aspect ratio (length divided by thickness) not exceeding 4, and a maximum surface area of 101 in.^{2} (0.065 m^2).

Table 3.1 Strength recommendation according to IS:15658:2006

Grade designation of paver blocks	Specified characteristic compressive strength of paver blocks at 28 days (N/mm^{-2})	Traffic category	Recommended minimum paver block thickness (mm)
M 25–M 30	25–30	Non-traffic	50
M 30–M 35	30–35	Light	60
M 35–M 45	35–45	Medium	60, 80
M 45–M 55	45–55	Heavy to very heavy	80, 100, 120

W. Mampearachchi, *Handbook on Concrete Block Paving*,
https://doi.org/10.1007/978-981-13-8417-2_3

South African standard SANS 1058–2006 (SABS, 2006), which was originally based on a compressive strength measurement and was later modified to SANS 1058–2012 (SABS, 2012) where two new performance measurement techniques were included: tensile splitting and abrasion testing. Yet, SANS 1058–2012 still refers to the SANS 1058–2006 compressive strength requirements to help the industry better understand the new performance measurements. For example, Class 30/2.0, officially rated at 2.0 MPa (tensile strength), indicates the minimum compressive strength should be 30 MPa, and similarly, Class 40/2.6, officially rated at 2.6 MPa (tensile strength), recommends a minimum compressive strength of 40 MPa. Class 30/2.0 in SANS 1058–2012 substitutes the Class 25 of SANS 1058–2006, which was based on a compressive strength rating of 25 MPa, and Class 40/2.6 substitutes the Class 35 of SANS 1058–2006 based on a compressive strength rating of 35 MPa.

The British standard BS 6717-1:1993 (British Standards Institution, n.d.) specifies the minimum compressive strength of paving blocks, as 49 N/mm^2 and the minimum crushing strength of any individual block as 40 N/mm^2.

Hence, it is very important to adhere to a proper mix design procedure to produce paver blocks in compliance with standard specifications. Not only the strength, but also a number of factors during production and utilization of the blocks should be accounted when determining the optimum mixture proportions and water content. Therefore, it is important to understand the manufacturing process to develop or improve the mixture design for ICBP.

At an industrial scale, the blocks are cast using hydraulic or electrical machines. The ingredients are mixed thoroughly by a pan mixture and the mixture is then conveyed through a conveyor belt to the hopper. Once the hopper is full, the mixture is poured to the mold by opening the bottom window of the hopper. The mixture in the mold is subjected to vibration and compression to cast the blocks. Typically, a semi dry concrete mix is used to cast interlocking paving blocks. 12 mm coarse aggregate and a suitable fine aggregate (River sand, quarry dust, manufactured sand, sea sand, recycled aggregates etc.) is mixed together with ordinary Portland cement (OPC) to produce the concrete mixture. Pigments are mixed to the mixture to achieve required colors.

3.2 Factors Governing the Mix Design

3.2.1 Strength

Strength is one of the key parameters that evaluate the performance of ICBP. Hence the mixture design should mostly focus on strength. Since the water content is kept at a minimum level, strength can be mainly improved by optimizing the aggregates. The production involves heavy duty vibration, thus higher strengths can be achieved by selecting aggregates in a way in which the packing of the mixture is maximized.

3.2.2 *Water Content*

Rapid manufacturing using hydraulic or electrical machineries require the mold to be stripped off from the fresh blocks soon after it is cast. To facilitate this requirement, zero slump concrete is used to produce the blocks. The water content should be maintained at an optimum level to produce high quality blocks. Excessive water can produce slumpy blocks which will collapse soon after the mold is stripped off, whereas less water will produce very dry blocks which will crack soon after the removal of the mold. On the other hand, maintaining a lower water content ensures high strengths according to Abram's law.

3.2.3 *Surface Texture of the Blocks*

The aesthetic appearance is one of the main features of paver blocks especially for landscaping applications. Similarly, surface texture should be maintained at a satisfactory level to facilitate a smooth and comfortable ride while maintaining a good traction especially for applications involving vehicular traffic such as roads, car parks and terminals. The fine aggregate percentage of the mixture together with water content governs the level of surface texture of the block. Hence, selection of fine aggregate proportions should be done carefully to satisfy both the performance criteria as well as the aesthetic appearance of the block.

3.3 Mixture Design Methods

3.3.1 *Trial and Error Method*

This method is the most primitive yet widely used method among industrial block manufacturers. They initially use the most common mix ratios such as the cement to aggregate ratio of 1:8 and then adjust this mixture to achieve required strengths. Typically, the water content is mainly based on intuition and experience. Hence, inconsistency is visible in the strength of the blocks. High strength variations within the same batch of blocks can be seen due to the improper mixture design. Segregation of aggregates is also a major problem of this mixture design method.

3.3.2 *ACI Method*

American concrete institute has specified a standard method to determine suitable mixture proportions for concrete. To facilitate a zero slump concrete mixture design, a special annex has been introduced. However, the use of ACI method does not provide a sustainable mixture design, as it uses a high amount of cement paste to gain higher strength classes. According to Shahriar, Amin, Ahmad, and Hossain (2017) the main reason for inconsistencies in the ACI mix design method is because of not accounting for the aggregate surface area when determining the cement paste (Shahriar et al., 2005).

3.3.3 DOE Method

The British method of concrete mix design, commonly referred to as the "DOE method", is used in the United Kingdom and many other parts of the world and has a long-established record. DOE method also specifies a method based on strength and durability of the concrete. However, according to Kumbhar and Murnal (2012) the DOE method also has several limitations. The higher fine aggregate content given by the DOE method makes sandy mixtures and this fine aggregate content cannot be changed for specific cement contents. One of the major concerns of the DOE method is that it does not have a specific method of combining quantities of different sized aggregates (Kumbhar & Murnal, 2012).

3.3.4 Particle Optimization Method

Particle optimization method is one of the most advanced methods that can be used to produce high performance concrete mixtures with required properties. The method involves a careful selection of aggregated proportions so that the packing density of the mixture can be increased. Higher packing density leads to a lower void content. This results in a lesser amount of cement usage. Also, having the full control of the voids percentage and aggregate mixture proportions makes this method very useful for specific applications like ICBP (Hettiarachchi & Mampearachchi, 2018, 2019). However, the determination of the aggregate content and mixture proportions using the packing density method is not widely used due to the lack of guidelines and information. There is no proper methodology available to follow on developing an optimized mixture for the use of ICBP manufacturing.

3.4 Problems Faced by the Industry Related to Mix Designs

Industrial block manufacturers face many problems due to the lack of knowledge on mix design concepts and lack of guidelines on mix design for block pavers. Variations in strength within the same batch of blocks, high usage of cement, high usage of fine particles in the mix, cracking of blocks soon after the removal of the mold, deformation of block shape due to slump, poor surface texture are some of the major issues existing in the industry. Following are the main factors that cause these issues.

3.4.1 Improper Aggregate Gradations

A field survey was carried out to understand the industrial ICBP manufacturing procedure. Figure 3.1 shows the 0.45 power curve for aggregate gradation used for various grades of concrete used by the selected manufacturer with standard limits proposed by South African (SA) standards (SANS1058, 2012). 0.45 power maximum density line provides a guide to blend aggregates to get the maximum density. Similarly, the Indian standard (IS15658, 2006) proposes a coarse aggregate to fine aggregate mix of 60:40 for ICBP manufacturing. As shown in Fig. 3.1, the South African standard specified limits are also closer to the maximum density line. It is evident that the industrial aggregate mix is not within the minimum and maximum lines specified by SA standards (SANS1058, 2012) as well as distant to the maximum density line. The high percentage (35–45%) passing the 0.075 mm sieve is also a major concern. These mixes may have a high amount of voids (40%) which will eventually result in either low strength or high cement paste consumption.

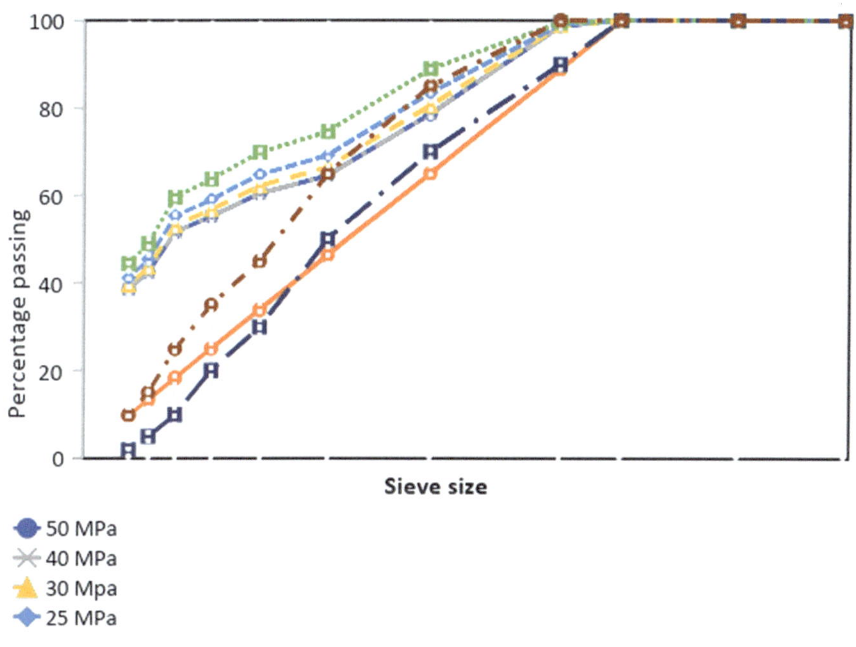

Fig. 3.1 0.45 power curve and aggregate size distribution at an industrial block manufacturer

3.4.2 Water/Cement Ratio

Water content is another major factor that affects the quality of the blocks. From the field survey, it was observed that many block manufacturers do not have a standard method to apply the amount of water to the mixture. Instead, they use intuition of an experienced person to add water to the mix. For example, water is added to the mixture until the mixture is not too dry and not too wet. A shovel is put to the mixture and removed. Then the surface of the shovel is observed. When the surface of the shovel has ripple marks due to water and cement, it is identified as the optimum water content for the block production. Due to this reason, the water cement ratio is not consistent throughout the production. Each mix has a different water cement ratio. This leads to inconsistent block strengths.

3.4.3 Improper Curing Methods

Another major factor that causes strength and durability problems in the blocks is improper curing. It was observed that the selected manufacturers do not apply proper curing methods. The blocks are stacked over each other, layer by layer and the water is poured using a garden hose to each stack of blocks. This method does not cure the blocks in the middle of the stack. Hence, those blocks can be subjected to strength and durability issues. Another substandard practice of the manufacturers is exposing the fresh blocks directly to heavy sunlight. This causes the immediate evaporation of the water that is dedicated to the hydration process. Hence, all the cement in the mixture may not hydrate since the mixture is using a very low amount of water to maintain the zero-slump characteristic of the concrete.

3.5 A New Mixture Design Method for the ICBP Using Packing Optimization

To overcome these issues, a proper mix design methodology based on the industrial block manufacturing process and application should be developed. The particle optimization method can be effectively utilized to determine the suitable mixture proportions for ICBP. The method can greatly reduce the amount of cement usage while maintaining the strength, surface texture and other required properties mentioned in Sect. 3.2. This section describes the methodology in detail. The mixture design methodology involves several steps and tests.

3.5.1 Step 1—Tests on Aggregates

In order to calculate the packing density of the mixture, it is important to identify the properties of aggregates. The following tests need to be done as the first step.

(i) Shape measurement

Shape of aggregates can be measured using a Vernier caliper. The length, breadth and width of the particles are measured as shown in Fig. 3.2, and the shape factor is calculated by Eq. 3.1.

$$F = \frac{\left(\frac{W}{L}\right)}{\left(\frac{T}{W}\right)} = \frac{LT}{W^2} \tag{3.1}$$

(ii) Surface texture measurement

To measure the packing density of aggregate mixture, the surface texture of the dominant aggregate particle should be known. Surface texture is measured using the British pendulum test (Fig. 3.3). The British pendulum number of aggregate surface is taken as the surface texture value.

(iii) Specific gravity measurement

It is important to know the specific gravity of the aggregates used in the mix design. Since packing density measurement is a volumetric property, the mixture proportions are obtained in volumes. Hence, to convert the volumes to weight it is important to measure the specific gravity of the aggregates. The specific gravity of aggregates can be measured using standard AASHTO T 85 and ASTM C 127 tests.

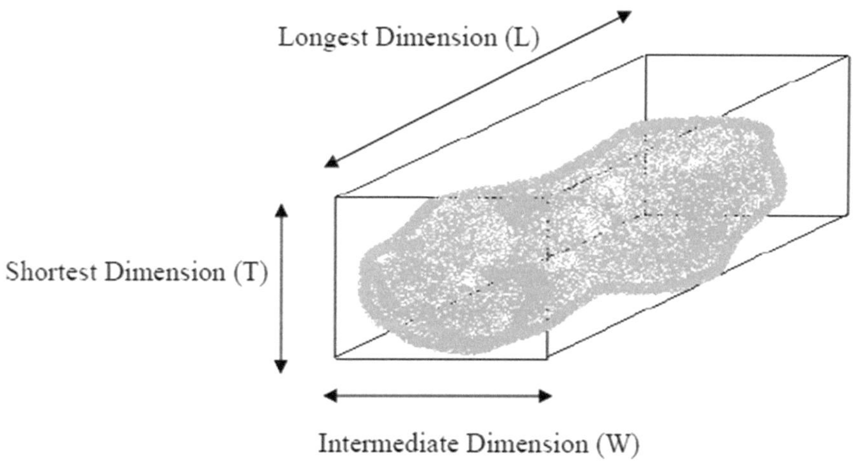

Fig. 3.2 Dimensions of a particle

Fig. 3.3 British pendulum tester

3.5.2 Step 2—Selection of Vibration Frequency

Vibration frequency is one of the main factors that govern the level of compaction of concrete mixtures. According to Hettiarachchi and Mampearachchi (2018), there is an optimum vibration frequency that provides the maximum packing density in a mixture. Hence, it is important to know the vibration frequency that can be exerted by the particular block casting machine in order to determine the packing density of the mixture. The vibration frequency may differ from machine to machine, and some sophisticated machines can change the frequency to the required levels. Therefore, the model requires the applicable vibration frequency to predict the packing density of the mixture.

3.5.3 Step 3—Determination of Packing Density (Model Application)

A packing model has been developed to predict the packing density of complex aggregate mixtures. This model can be simply used to determine the packing density of binary particulate mixtures. The model is given by Eqs. 3.2–3.11.

$$\frac{1}{\emptyset_i^*} = \left(\frac{r_i}{\emptyset_i} + \frac{r_j}{\emptyset_j}\right) - (1-b)\left(1-\emptyset_j\right)\frac{r_j}{\emptyset_j}\left[1 - c\left(2.6^{r_j} - 1\right)\right] \tag{3.2}$$

$$\frac{1}{\emptyset_j^*} = \left(\frac{r_i}{\emptyset_i} + \frac{r_j}{\emptyset_j}\right) - (1-a)\frac{r_i}{\emptyset_i}\left[1 - c\left(3.8^{r_i} - 1\right)\right] \quad (3.3)$$

$$\emptyset_{mix} = min(\emptyset_i^*, \emptyset_j^*) \quad (3.4)$$

- Size class i represents small particles
- Size class j represents large particles
- r_i and r_j: volumetric fractions of solids in size classes i and j; $(r_i + r_j = 1)$
- $\emptyset_i$ and $\emptyset_j$: packing densities of size class i and j
- $\emptyset_i^*$: Packing density of the mixture when size class i is dominant
- $\emptyset_j^*$: Packing density of the mixture when size class j is dominant

where a, b and c are loosening effect, wall effect and wedging effect parameters

$$a = 1 - (1-s)^{1.5} - 3 \times s \times (1-s)^W \quad (3.5)$$

$$b = 1 - (1-s)^X - Y \times s \times (1-s)^Z \quad (3.6)$$

$$c = 0.36.\tanh(12.s) \quad (3.7)$$

where

$$W = -3.45.\gamma + 5.85 \quad (3.8)$$

$$X = 0.1 \times \beta + 4 \quad (3.9)$$

$$Y = -0.0016.\omega^2 + 0.3536.\omega - 0.2308 \quad (3.10)$$

$$Z = 0.0008 \times \omega^2 - 0.177 \times \omega + 14.1 \quad (3.11)$$

The following details needs to be given to the model

S Size ratio (Diameter of smaller particle/Diameter of large particle)
γ Shape factor
β Surface texture (BPN)
ω Vibration frequency (Rad/s).

3.5.4 Step 4—Water/Cement Ratio Calculation

Water/cement ratio is critical in the ICBP concrete mix. Since a dry concrete mix is used to cast blocks, the water/cement ratio needs to be kept to an optimum level. Nevertheless, too dry a mix will break the fresh block after casting, as well as the required surface texture cannot be achieved. On the other hand, too wet a concrete

(a) 0.3 W/C (b) 0.35 W/C (c) 0.4 W/C

Fig. 3.4 Results of box test for different water/cement ratios

mix will slump the block, casting and stripping of mold afterwards is difficult due to a sticky concrete mix. Therefore, to select the most suitable W/C ratio, a box test needs to be carried out (Cook, Ley, & Ghaeezadah, 2014). Typically, 0.34–0.38 W/C ratio provides a better concrete mix which satisfies the required criteria. Figure 3.4a–c show the surface texture resulted in box tests with respect to the W/C ratio of 0.3, 0.35 and 0.4 respectively.

3.5.5 *Step 5—Strength Class Determination (Excessive Cement Paste)*

Excessive cement paste is the amount of cement paste that will remain after filling all the voids in the aggregate matrix. Excessive cement paste will coat the aggregate and act as a bonding agent for the aggregates. Increase of the excessive cement paste will increase the thickness of the bond between two aggregates, thus improving the strength of the bond. Since the aggregate is packed as densely as possible and the water/cement ratio is kept to an optimum level, it is the excessive cement paste that will improve the strength of the concrete. According to Taylor, Bektas, Yurdakul, and Ceylan (2012) excessive cement paste can be increased up to 130% to increase the strength of the concrete (Taylor et al., 2012). As shown in Table 3.2, when the excessive cement paste is increased the strength of the concrete increases.

Table 3.2 Effect of cement paste on strength

Excessive cement paste (%)	28 days strength (N/mm^2)
10	15–20
15	20–25
20	25–30
25	30–35
30	35–40
35	40–45
40	45–55

3.5.6 Step 6—Calculation of Mixture Proportions

Once the packing density is determined and excessive cement paste is identified, the mixture proportions can be calculated. The following is a sample calculation of a mixture proportion using the developed method.

Total maximum packing density of the mix (From the model) = 0.75
Volumetric fraction of coarse aggregates at maximum packing = 0.7
Voids content = 1 − 0.75 = 0.25
Paste content 10% in excess of voids = 0.25 + 0.1 * 0.25 = 0.275
Volume of aggregates = 1 − 0.275 = 0.725 m^3
Volume of coarse aggregates = 0.7 * 0.725 = 0.5075 m^3
Weight of coarse aggregates = 0.5075 * 2.6 * 1000 = 1319.5 kg/m^3
Volume of fine aggregates = 0.3 * 0.725 = 0.2175 m^3
Weight of fine aggregates = 0.2175 * 2.65 * 1000 = 576.375 kg/m^3
Water Cement ratio (From box test) = 0.35
Total Paste = Cement + water = $C/3.15 + 0.35C = 0.667C = 0.275$
Cement content = 0.275 × 1000/0.667 = 412.3 kg/m^3
Water Content = 0.35 × 412.3 = 144.3 kg/m^3

3.6 Summary of the Mix Design

See Fig. 3.5.

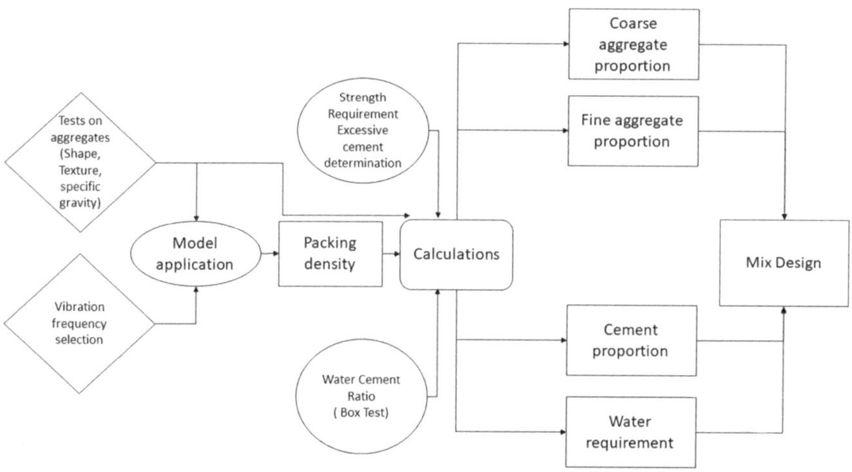

Fig. 3.5 Mix design methodology flow chart

Table 3.3 Concrete mix proportions and strengths

Mix design methodology	Fine (kg/m^3)	Coarse (kg/m^3)	Cement[a] (kg/m^3)	Water (kg/m^3)	7-day strength (kN/m^2)	28-day strength (kN/m^2)	Cement reduction[b] (%)
Proposed methodology	850	364	735	262	32	43	27.9
Three parameter model	432.5	649	870	304.5	33	45	14.7
Toufar model	644	346.5	910	318.5	31	42	10.7
Compressible packing model	555.5	370	950	333	33	41	6.8
ACI method (G40)	575	910	840	290	27.15	40.7	17.6
Industrial mix	1310	340	1020	N/A	30	45	–

[a]35% excessive cement content
[b]Reduction of cement is calculated as a percentage compared to the industrial mix

3.7 Case Study

The developed mix design methodology was tested in the laboratory and the blocks were cast using industrial block manufacturing machines. The comparison of mix design with several other standard mix designs, are given in Table 3.3.

The proposed methodology, the three-parameter model, the Toufar model and the compressible packing model are based on the particle optimization method. The mix proportion of large particles (coarse) to smaller particles (fine) which yield the maximum packing density was selected for the concrete mix design. The excessive cement paste was selected as 35% for the case study. Water cement ratio was selected as 0.35 for trial mixes using the box test and unconfined compressive strength was performed to determine the strength. The results show that the proposed methodology uses a minimum amount of cement to achieve the required strength. The results are given in Table 3.3. Figure 3.6 shows the fresh blocks cast from the proposed methodology. As visible in the Fig. 3.6 the blocks were having a very good green strength to hold its shape and the mixture was suitable to maintain fine details of the texture.

Table 3.4 shows the various block strengths obtained from the proposed mix design methodology for different excessive cement contents.

3.8 Advantages and Limitations of the Proposed Method

The proposed method is developed specifically based on the ICBP manufacturing process and application aspects. Hence, the concrete mixtures batch from this method

Fig. 3.6 Sample ICBP using optimized concrete mixture

Table 3.4 Strength gain for different excessive cement content

Sample	Excessive cement content %	3-day strength (N/mm^2)	7-day strength (N/mm^2)	28-day strength (N/mm^2)
1	20	12.4	18.9	26.3
2	25	15.2	21.4	30.5
3	30	18.6	27.6	36.4
4	35	22.6	30.1	43
5	40	28.3	37.4	52.1

will produce high quality blocks. The ability to input vibration aspects, aggregate surface texture and shape characteristics will accurately model the aggregate packing characteristics. Hence, mixtures with higher packing densities can be obtained. This will produce better mixtures that have a minimum amount of voids. Lesser voids require a lesser amount of cement paste to fill the voids. Thus, the mixtures developed from this method will use a lesser amount of cement. Cement is the most costly material in a concrete mixture. The lesser the cement content the lesser the production cost. Further, this method will have greater profit margins for manufacturers and reduced unit prices for customers. Also, this method optimizes particle packing taking the vibration of the machine in to account, the mixture will use less energy to compact, reducing the energy wastage.

However, there are some inherent limitations of this method. The method involves multiple tests for aggregates as well as for the water content. Also, the model application is slightly complicated. Therefore, a trained personal is needed to perform the mix design. Another limitation of this method is that it is too specific for a given machine and for given aggregates. Hence, if the source of the aggregate is changed or a different machine is utilized, the mix design needs to be repeated with new parameters.

Chapter 4
Concrete Block Pavement Design

4.1 Pavement Design Method for CBP

There are so many barriers to the widespread use of concrete block paving. This can be attributed to the scarcity of an experienced workforce, and the lack of reliance in design methods that are relevant to local conditions. Consequently, there is a widely perceived need to establish a sound technological basis for the design and construction of concrete block paving.

According to the literature, there are four design concepts mainly identified for concrete block paving: equivalent design concept, catalog design method, research based design method and mechanistic analysis (Concrete Manufacturing Association, 2004). The recent study of Mampearachchi and Gunaratne (2010) was conducted using a prototype model; hence this can be categorized as a research based design approach. Results generated after testing the prototype model were used to create a Finite Element Model (FEM) using **SAP2000 Finite Element software**. Hence, this is a combination of mechanistic and research based design concepts. A developed design concept was used to build up design charts by considering the deflection measurement of the different support conditions.

This chapter focuses on the development of design charts, and on the introduction of appropriate support conditions for low volume roads.

4.1.1 Development of Design Charts

Design charts were introduced for different subgrade, sub base and base conditions. The support conditions used are listed below.

Subgrade CBR values	5, 10, 15, 20 and 30%
Sub base CBR value	30% (Type 1 soil)
Base CBR value	90% (Aggregate Base Course).

W. Mampearachchi, *Handbook on Concrete Block Paving*,
https://doi.org/10.1007/978-981-13-8417-2_4

Design charts were developed using a software model and the following dimensions were used.

Dimension of Concrete Blocks	225 mm × 100 mm × 75 mm
Sub grade thickness	2000 mm
Road width	3.1 mm (Minimum lane width) (Fig. 4.1).

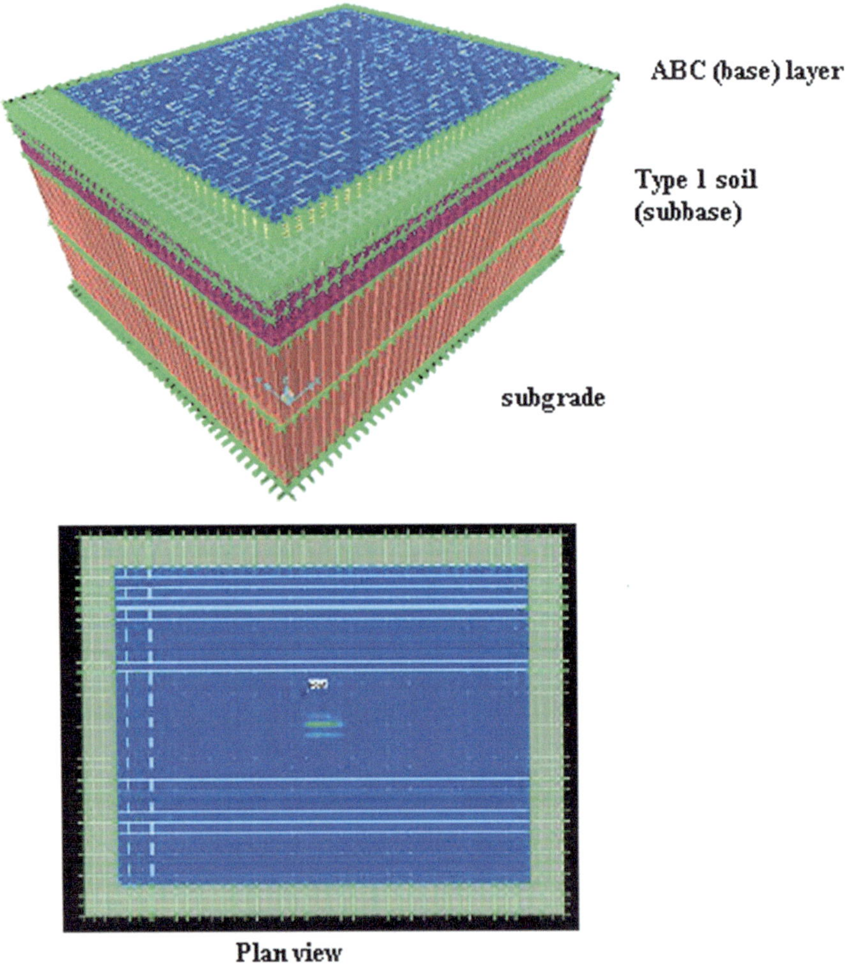

Fig. 4.1 3-D view and plan view of developed software model

4.1.2 Design Chart I (Base Improvement with 150 mm ABC Layer)

Design chart I (Fig. 4.2) was introduced based on Sri Lankan practices (Base improvement with 150 mm thick ABC layer and Type 1 soil as subbase). Base layer thicknesses of most roads constructed have been restricted to a maximum of 150 mm due to the high material cost. Special circumstances requiring higher thicknesses are achieved by adding type 1 soil as an improvement layer.

Subgrade conditions were improved starting from 5% of CBR value and were increased as 10%, 15% and 20%. Then subgrade was improved with a 150 mm of base layer and different subbase layer thicknesses (Fig. 4.1). The deflections were measured under an applied pressure (almost 100 psi of maximum allowable tire pressure).

4.1.3 Design Chart II (Base Improvements Without Sub Base Layer)

Design chart II (Fig. 4.3) was developed for ground improvement with the base layer (ABC) and was designed for different base layer thicknesses.

Based on chart I and II, suitable subgrade improvements for different ground conditions can be determined. The charts show the deflections curves for various

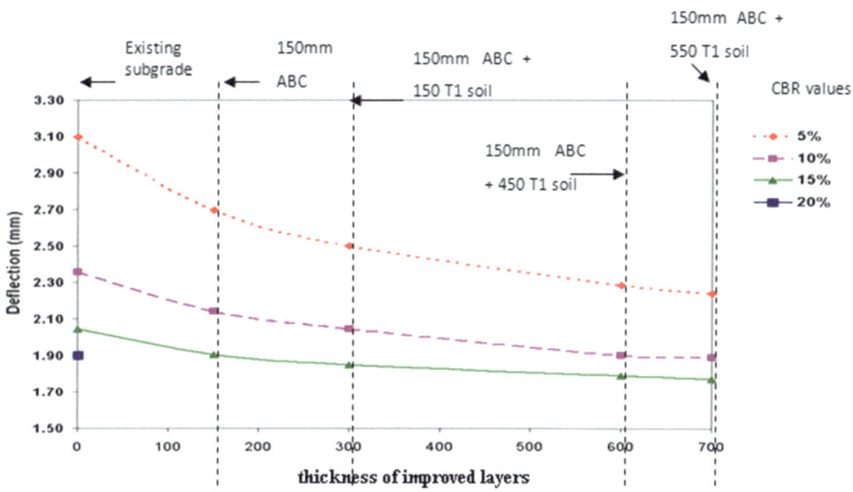

Fig. 4.2 Design chart I (different support condition vs. deflections). *Note* Thicknesses of the improved layers are mentioned as follows. E.g: 150 mm ABC +150 mm T1 soil—150 mm thick aggregate base layer and 150 mm type 1 soil layer used as an improved layer

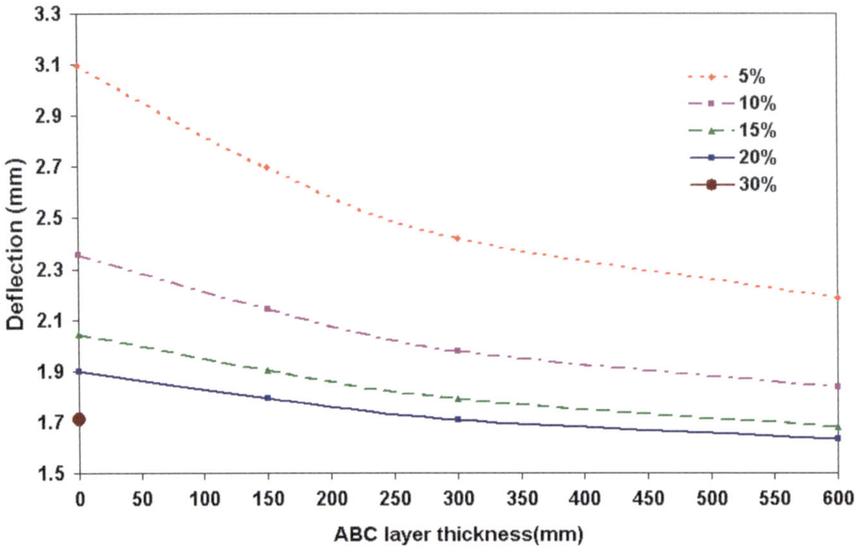

Fig. 4.3 Design chart II (different support condition vs. deflections)

subgrade conditions (CBR 5–20) and selected base improvements. New design guidelines can be introduced for low volume roads based on the design charts. Nevertheless, there should be a reference deflection value for the optimum subgrade condition in particular applications. Therefore the optimum support condition is obtained based on the existing design methods and field surveys, which will be described in Sect. 4.2.

4.2 Optimum Support Condition for Low Volume Roads

Field survey

There are only a few concrete block paved roads in Sri Lanka. Hence, the choices were restricted when collecting data for different ground conditions. Usually low volume roads are subjected to an Average Daily Traffic (ADT) of around 300 vehicles per day (vpd). RDA pavement design guidelines refer to roads subjected to over 0.3 million standard axles (msa) within the design life (10 years) as low volume roads. Thus, selected concrete block paved low volume roads were evaluated considering the following performance criteria (Table 4.1).

- Service period
- Length of the road
- Width of the road
- Usage of road
- Dimension and Quality of the Concrete Block

Table 4.1 Observed data from concrete block paved roads

	Bandigoda road	Piduruwella road	Munamaldeniya road
Service period	1 year, 8 months	2 years, 9 months	2 years
Length of the road	150 m	280 m	150 m
Width of the road	3 m	3 m	3 m
Usage of road	Access to village	Access to village	Access to temple
Dimension of concrete block	225 mm × 100 mm × 75 mm	225 mm × 100 mm × 75 mm	225 mm × 100 mm × 75mm
Quality of the concrete block (0–10)	8	6	8
Flatness (Sect. 2.6.1)	<5mm	<8 mm	>10 mm
Alignment (Sect. 2.6.1)	Not satisfied	Not satisfied	Not satisfied
CBR values in subgrade	Over 35%	25%	25% (150 m) 11%

Note None of the roads used a base layer (ABC) as an improvement layer

- Flatness (level difference between adjacent blocks measured using a straight edge)
- Alignment (straightness of joints)
- CBR values of support layer (CBR of subgrade or improved subgrade measured using DCP test).

Bandigoda and Piduruwella roads performed well under existing subgrade conditions and the Munamaldeniya road did not satisfy the flatness criteria. Therefore, it is concluded that Munamaldeniya road does not have an adequate support condition for existing traffic volumes. Piduruwella road satisfied the criteria for low subgrade strength (25% CBR) condition when compared to other roads. Thus 25% CBR value can be used as a sufficient support condition for low volume road conditions, based on the field survey (Fig. 4.4).

Based on the above field survey, it can be stated that Piduruwella road and similar road projects in the Western province behaved well with the subgrade condition of 25% CBR value. Hence, according to the field observations, the subgrade strength of 25% CBR can be considered as a safer condition, and it needs no improvement for the concrete block laying work. From chart II, 125 mm ABC can be indicated as an improvement layer for the 20% CBR subgrade condition, equivalent to 25% CBR subgrade condition. Simulated model of the Munamaldeniya road, subjected to 7.174×10^{-4} KN/mm^2 pressure gave a maximum deflection of 2.22 mm, which is the equivalent deflection of 10% CBR subgrade condition, with an improved 100 mm ABC layer under the same loading (reference to chart II). Therefore, the subgrade condition of 10% CBR with a 100 mm improvement layer can be considered as a situation of failure according to field observations.

Fig. 4.4 Field survey

Figure 4.5 represents with clarity a comparison between existing design methods and field observations. It clearly shows the failure condition and safe condition of low volume roads, and a comparison between design methods. Some of the findings can be summarized as below.

There is no need for any improvement layer for a subgrade condition above 25% CBR.

If the subgrade condition is 20% CBR with an improvement layer of 150 mm ABC, it would be considered safer.

Subgrade condition of 10% CBR should have an improvement ABC layer more than 600 mm in order to achieve a safe condition.

According to the chart, Lock pave design is not adequate for 10% CBR subgrade condition to be safer (even 80 mm thick block).

Therefore, it can be observed that there should be a particular subgrade strength that does not need any improvement layer for concrete block laying. According to the Catalog method used in South Africa (Appendix A), a minimum subgrade strength of 15% CBR value can be used for low volume roads with a minimum thickness of 100–150 mm (minimum CBR of 45%) improvement layer. According to a field survey, 300 mm of ABC improvement layer is needed for the 15% CBR subgrade condition (Fig. 4.5), to achieve safer conditions. Therefore, it is recommended to use the subgrade strength of 15% CBR with an improvement layer, for low volume

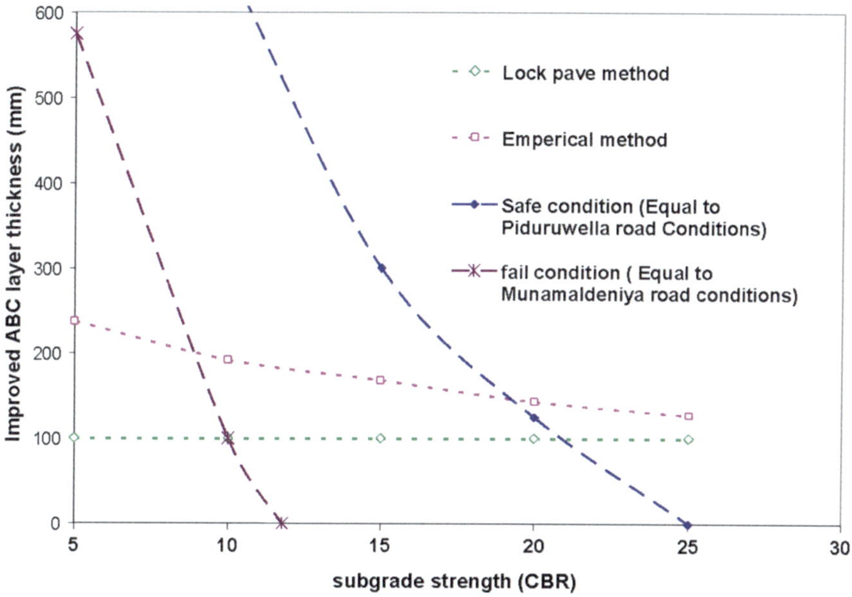

Fig. 4.5 Comparison between field survey and existing design method

traffic conditions. Figure 4.5 shows the proposed ABC thicknesses in available design guidelines, and safe and failure conditions according to field observations.

An improvement layer of 100 mm thickness is proposed by some guidelines for the 20% CBR (Tsohos & Iliou, 1994) condition, as a conservative approach. According to the study of Abril, Irastotza, and Josa (1994) design manual, if the average annual daily heavy vehicle traffic (AADHT) is between 20 and 49, there is no need for subgrade improvement for CBR values higher than 20% (Appendix B). The Sri Lankan Road Development Authority design also stipulates that there is no need for base improvements for concrete block laying, which has a subgrade strength above 20% CBR (Appendix C).

Therefore, considering the above findings, it can be proposed that concrete block laying work can be executed directly on low volume roads, which have CBR values above 20%. It is important to note that an accurate estimation of the CBR value in the safe condition could be obtained with a better sample of field observations, using more block paved roads. The lack of readily available block paved roads to conduct the field surveys within the stipulated time period of the study was a major obstacle for pooling a sufficient amount of surveillance data. Therefore, the proposed new design uses 20% (CBR), as an optimum subgrade strength for low volume roads. Figure 4.8, shows the proposed safe condition for low volume roads (depicted in Fig. 4.5). Design chart I and II are used to determine base improvements for weaker support conditions using the reference to deflection value (Figs. 4.6 and 4.7). Layer thicknesses extracted from Chart I and Chart II are shown in Table 4.2.

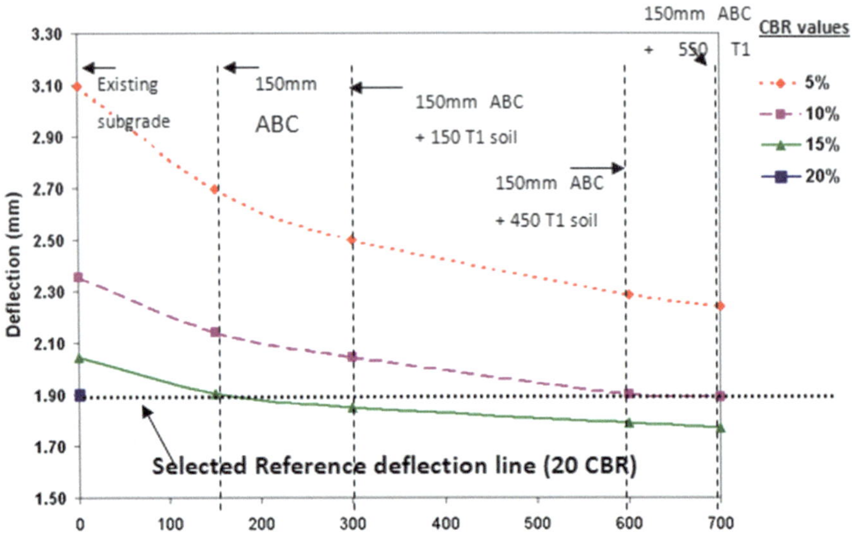

Fig. 4.6 Proposed design chart I for block paving with improved sub base

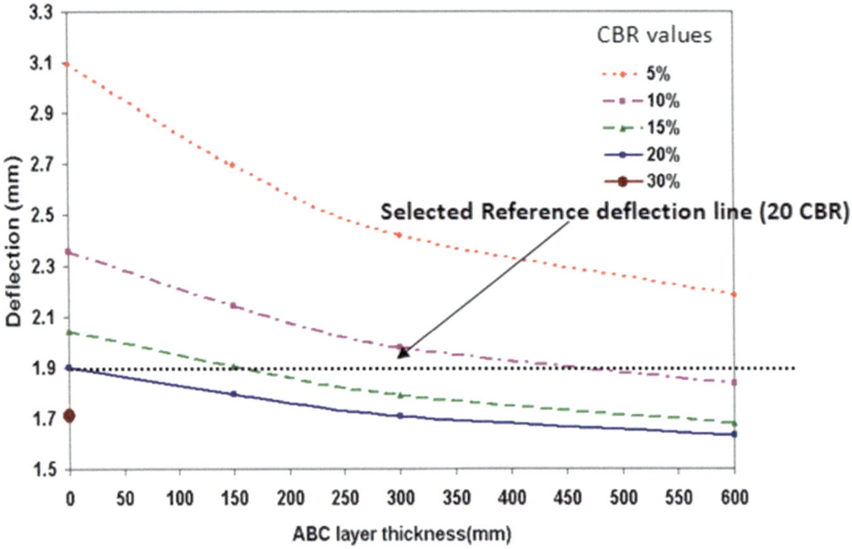

Fig. 4.7 Proposed design chart II for block paving with improved ABC layer

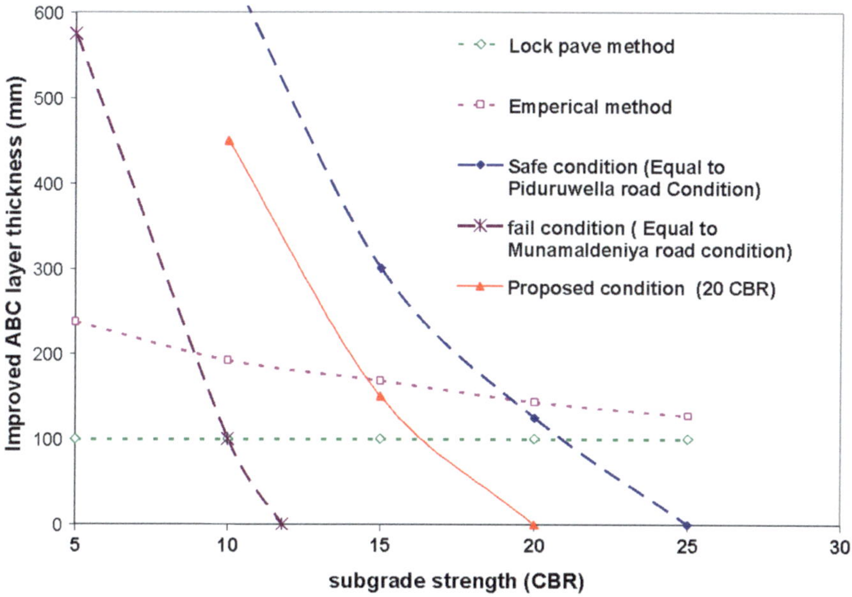

Fig. 4.8 Proposed design for CBP with improved sub grade

Table 4.2 Proposed designs for low volume roads

Subgrade condition (CBR)	Base layer (mm)	Sub base layer Type 1 soil (30CBR) (mm)
20	–	–
15	150	–
10	150	550
10	450	–

4.3 Proposed Design for Low Volume Roads

The design chart shows that to improve roads with subgrade CBR 10% or below, thicker base or subbase layers are needed. So, it is not economical to improve such subgrade conditions for block laying. It is recommended to introduce concrete block laying work for low volume roads which are having a subgrade strength higher than 10% (CBR).

Chapter 5
Good Practices in Concrete Block Laying

5.1 Design of Block Laying at Curves

It is necessary to cut the paving units into various sizes, in order to fit the edge restraints at curves. Rectangular blocks of similar or contrasting colours have been used as edging to minimize the visual effects of small errors in block cutting.

To avoid unsightly and potentially weak joints, it is often preferable to change the laying pattern at the curve. For example, as shown in Fig. 5.1, the curve itself can be installed in herringbone bond, yet the pavement can revert to stretcher bond at the approaches. Figure 5.1 shows the stretcher block laying pattern at approaches, and the laying pattern changes to herringbone at the tangents and curve. Cutting the blocks to meet the herringbone shape may be required at the curve.

5.2 Changes in Direction at Intersections and Corners

Changes in direction of a road pavement can be achieved by the use of special blocks. However, it is generally easier to choose blocks that can be installed in herringbone and simply cut the blocks to fit the edge restraints. Figure 5.2 shows the herringbone laying pattern at an intersection and a curve. At intersections, if a herringbone bond laying pattern is adopted, the paving can proceed without the need for construction joints.

When the aesthetic requirement of the shape of paving unit dictates, the stretcher bond can be used. Then only a 90° direction change in alignment can be achieved without cutting the blocks, as per Fig. 5.3.

The sequence of rectangular paving units between the main roadway and the side streets will permit different laying patterns to be used in two roadways. Figure 5.4, shows two different laying arrangements for the two roadways and joints, constructed with rectangular paving units.

W. Mampearachchi, *Handbook on Concrete Block Paving*,
https://doi.org/10.1007/978-981-13-8417-2_5

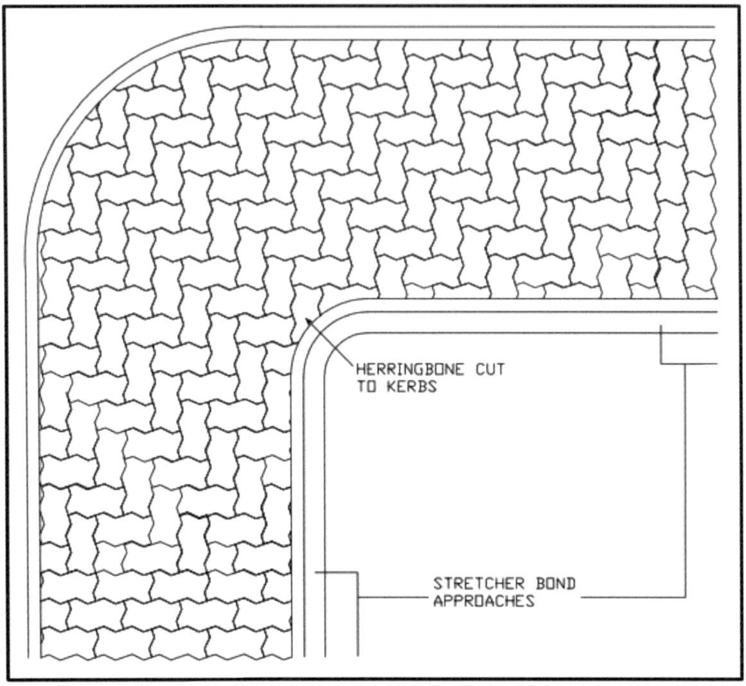

Fig. 5.1 Transition of stretcher laying pattern at curves

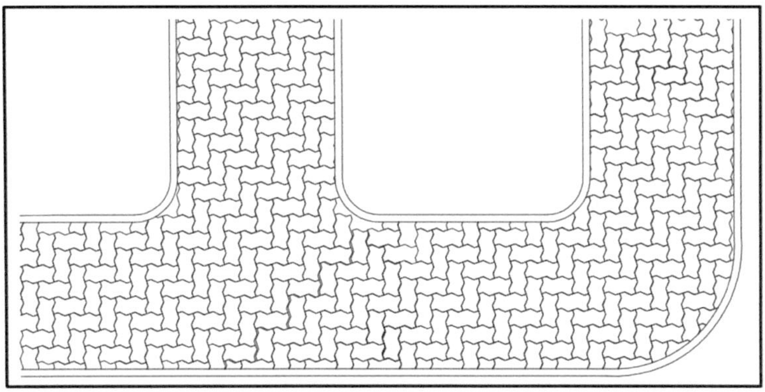

Fig. 5.2 Herringbone pattern at an intersection and a curve

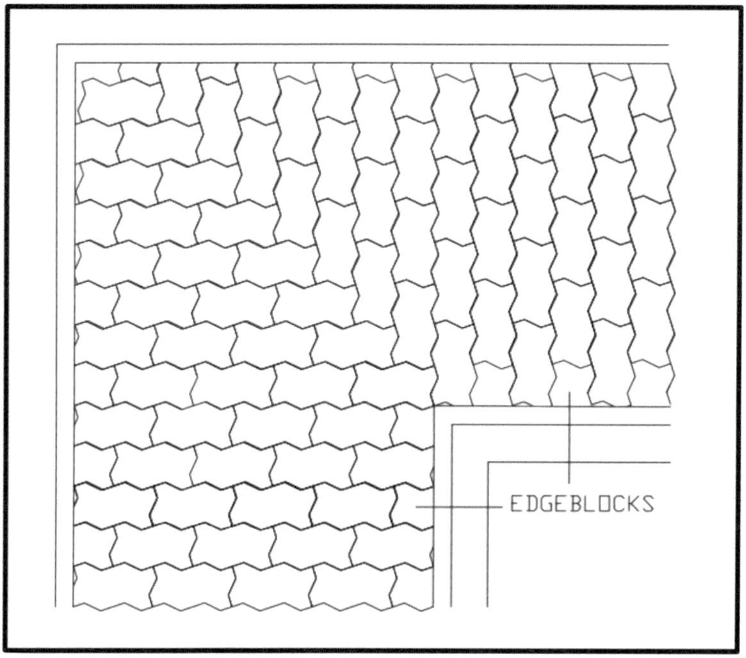

Fig. 5.3 Laying Pattern at 90° corner

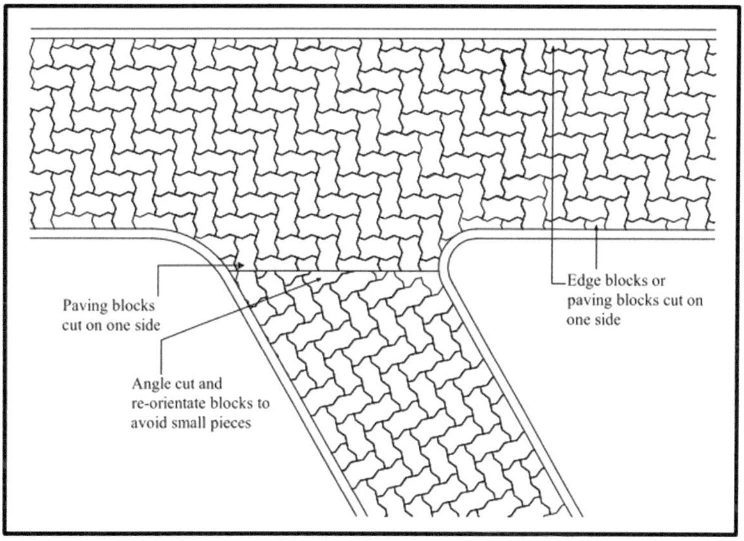

Fig. 5.4 Joints of two laying patterns at an intersection

5.3 Establishing the Laying Pattern

The laying patterns and faces should be established to permit fast and easy laying without the necessity of forcing a block between previously positioned blocks. Figure 5.5 shows the open laying faces. The blocks can be laid easily and several laying operators can work simultaneously. More working space is available since the laying has been started from an open space.

Figure 5.6, shows block laying with a close laying face. The block needs to be placed between previously positioned blocks.

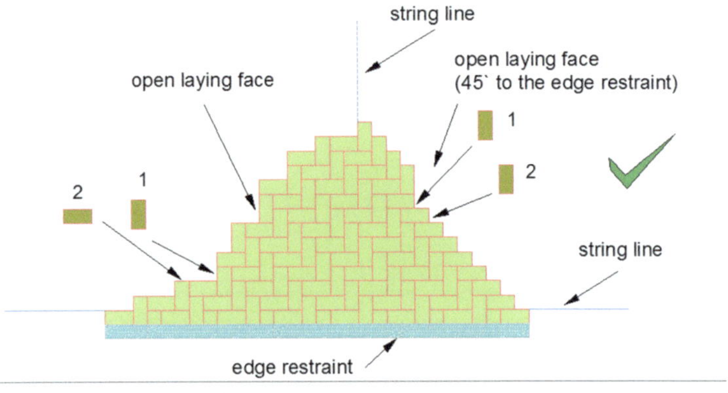

Fig. 5.5 Block laying with an open laying face

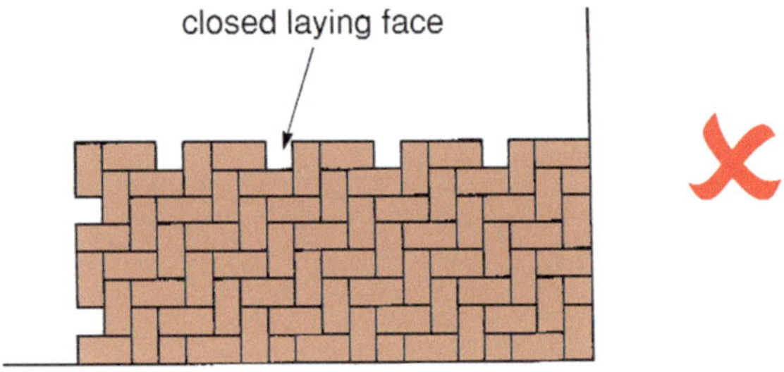

Fig. 5.6 Block laying with a close laying face

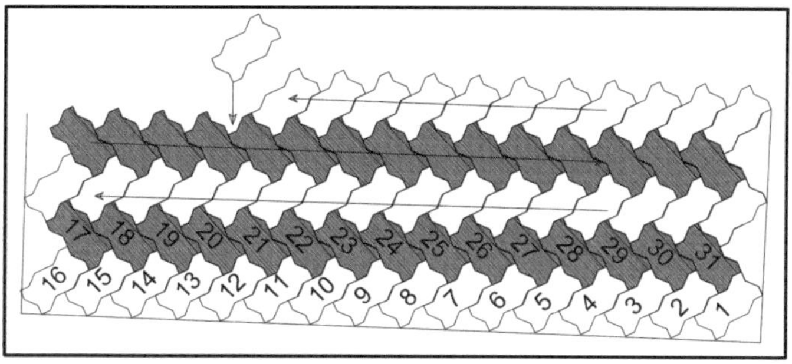

Fig. 5.7 Angle laying of Herringbone pattern

5.3.1 *Angle Laying (Block Axis Is Not Parallel to the Edge Restrain)*

The sequence of placing the block in angle laying is shown in Fig. 5.7. Only one operator can work in placing the blocks. Placing of blocks aligned to a string line is difficult with the angle laying of the herringbone pattern.

So, it is essential to maintain gaps (3–6 mm) between blocks using gauges in the first few square meters, until the operator is familiar with the laying pattern. The angle laying of the stretcher bond is shown in Fig. 5.8. A string line is used to align the blocks, and a uniform gap can be maintained between blocks. The effect of block size variation does not affect the alignment of blocks.

5.3.2 *Block Axis Is Parallel to the Edge Restrain*

The sequence of block laying in a herringbone pattern is shown in Fig. 5.9. Two or more operators can work independently in placing blocks. The direction of placing blocks is shown in Fig. 5.9. Gaps between blocks need to be maintained using gauges and string lines.

Figure 5.10 shows the placing of blocks according to the stretcher bond pattern. Block placing can be aligned to a string line.

Fig. 5.8 Angle laying of the stretcher bond pattern

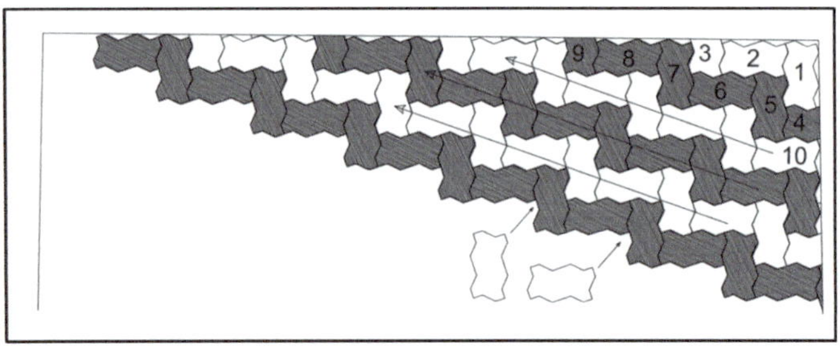

Fig. 5.9 Laying of the herringbone pattern to 90° or 0°

Fig. 5.10 Laying of stretcher bond pattern to 90° or 0°

Chapter 6
Maintenance of Concrete Block Pavement

Concrete block paving provides a hard surface which is aesthetically pleasing, comfortable to walk on, trafficable, durable and easy to maintain. Unlike the bituminous pavement, it is possible to complete the reinstatement work in concrete block paving with no visual evidence that a repair has been undertaken. Unlike the other type of pavements, the concrete block pavement gives the following advantages when maintenance work is undertaken.

1. Paving blocks can be lifted and re-laid without using machines that cause sound pollution. Hence, this is a low cost and environmentally friendly task. A paving block removal device is shown in Fig. 6.1.
2. After the reinstatement, the Block Paving Pavement can immediately be re-opened for public use.
3. Paving Blocks can be reused for other purposes, after being lifted from the pavement and hence reduces the wastage.

Fig. 6.1 Special block removal/extractor tool

W. Mampearachchi, *Handbook on Concrete Block Paving*,
https://doi.org/10.1007/978-981-13-8417-2_6

4. Block Paving allows sectional removal and reinstatement of hard standing areas using the original materials and blocks, without leaving an evidence of a repair.

6.1 Failure Categories

Maintenance work on concrete block paving can be undertaken at following situations.

6.1.1 Spalled Block Paving

The block pavers spalled as a result of the rectangular units having no space, and being installed in a tightly packed manner. The problem was exaggerated by the absence of an adequate base. Figure 6.2 shows the spalled pavers at the Bahrain Airport.

6.1.2 Failure Due to the Loss of Interlocking

Filling joints with sand is essential to maintain the horizontal and vertical interlocking of the blocks. Joint gap variation and settlement of blocks as shown in Fig. 6.3, is due to the lack of horizontal and vertical interlocking. Sand bed preparation and relaying of the blocks with joint sand is essential to maintain a uniform gap and avoid future settlements.

Fig. 6.2 Spalled pavers at Bahrain Airport

Fig. 6.3 Jointing material has been lost from pavers at leeds resulting in loss of interlock

6.1.3 Failure Due to Jointing Sand Being Washed or Vibrated into the Bedding Course

Figure 6.4, illustrates the concrete block pavement after several months of service. It can be seen that joint filling sand was being lost during the service. Loss of jointing sand can occur due to several reasons.

6.1.3.1 Improper Grading of the Jointing Sand

ASTM C 33 or Type 1 sand as per Indian standard should be selected for joint filling. Coarse sand can block the mouth of the joints, but the full depth of the joints may not

Fig. 6.4 Failure of surfacing on a car park deck as a result of jointing sand being washed/vibrated into bedding course

be filled. On the other hand, fine sand may not have proper frictional characteristics and may migrate to the bedding sand with time.

6.1.3.2 Wet Sand and Blocks

It is very important to maintain blocks and sand, in a dry condition to fill joints fully. Wet sand can stick to the mouth of the blocks and prevent the filling of the joints. Same thing would happen with wet blocks. So, the dry sand and the dry condition of blocks should perform better. If a wet climate prevails, sprinkling water during joint filling will help the sand to flow through the joint opening.

6.1.3.3 Insufficient Joint Width

The joint width should be selected according to the gradation of the sand. An insufficient joint width will not allow sand particles to move through the joints, and only the joint opening will be filled during construction (false joints).

6.1.3.4 Poor Compaction of the Sand Bed and Jointing Sand

Jointing sand will be lost if the sand bed is not properly compacted. Surface compaction using a plate vibrator or vibrating roller after spreading sand on blocks ensure the filling of all joints. Vibration will facilitate the flow of sand particles through the joints. It can be compacted well with an adequate roller compaction.

Relaying of concrete blocks with proper material for the sand bed and joint sand, joint spacing and the compaction of sand bed, and filling joints, need to be considered in repairing such failures.

6.1.4 Larger Gaps (Non-uniform) Between Interlocking Blocks

A joint gap should be maintained between 3 and 6 mm spacing during construction for the effective interlocking of the blocks and for the better appearance of the road. Figure 6.5 shows the non-uniform gaps in a CBP road. There are several reasons for non-uniform gaps in block paving; (1) the improper size of the blocks, (2) poor construction practices and (3) the improper compaction of joint filling sand.

Tolerance allowed for concrete block dimensions (width and length) is ± 2 mm and the alignment tolerance allowed is 15 mm for 15 m length. During construction, blocks should be placed aligned to a string line to maintain a uniform gap between blocks.

Relaying of blocks with a 3–5 mm joint spacing is recommended.

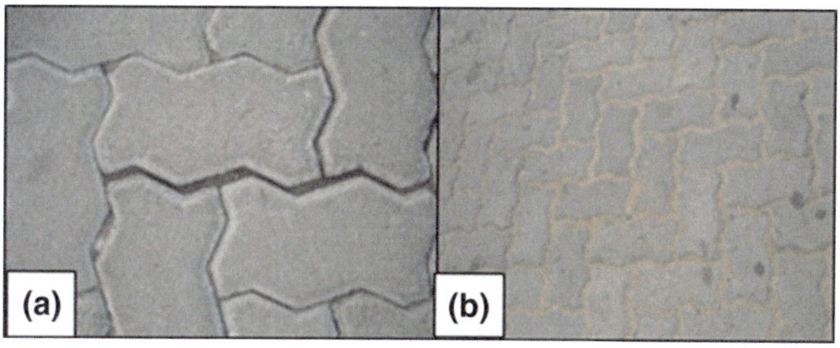

Fig. 6.5 **a** Non uniform gaps not filled with sand. **b** Non uniform gaps filled with sand

6.1.5 Sand Patches on the CBP Surface

The sand patch shown in Fig. 6.6 can be due to several reasons.

6.1.5.1 Poor Compaction of the Sand Bed and Support Layers

Proper compaction should be maintained in the prepared subgrade or subbase and base layers. Further, sand bed should be compacted under proper moisture conditions to avoid the sinking of concrete blocks.

Fig. 6.6 Sand patches on CBP surface

6.1.5.2 Poor Construction Practices in Block Laying

During block laying, proper quality control measures should be followed to avoid the surface becoming uneven in CBP roads. Only a 2 mm level difference is allowed between two paving units, and the flatness of the road should be checked. A tolerance of 10 mm is allowed for a 3 m straight edge.

Relaying of blocks at depressed areas after compacting and leveling the base and sand bed is recommended. Flatness should be checked during construction and after compaction to avoid depressed areas.

6.1.5.3 Loss of Sand in Gaps

Sand can be washed out from joints and the sand bed if water infiltrates through joints. Packing of joints with sand, compaction of the sand, maintenance of surface drainage (cross fall) are essential to minimize the infiltration of runoff water through joints. Figure 6.7 shows the loss of sand in the joints and the bed. Relaying of blocks taking adequate measures to avoid washout is important.

6.1.6 Washing Out of Blocks

Figures 6.8 and 6.9 show the washing out of cement from the surface. There are several reasons for the washing out of cement from pavers. They are:

I. High water/cement ratio (use of more water in the preparation of the concrete mix)
II. Lack of curing.

Fig. 6.7 Washout of sand from joints and bed

Fig. 6.8 Washout of blocks

Fig. 6.9 Wearing out of blocks

It is essential to cure concrete blocks at least 7 days in order to provide a durable surface. Figure 6.9 shows a worn out block among fine blocks. It is apparent that an effective curing method had not been used. Some blocks in a pile may not be subjected to curing, during the curing period. Figure 6.10 shows the water curing of a stack of blocks in a manufacturing yard. Some of the blocks are not being cured due to the ineffectiveness of the selected curing method. This will lead to the non-uniformity of the strengths of concrete blocks, and to the creation of defective blocks.

III. Low cement content

It was found that the strength of concrete blocks reduce as the cement to aggregate ratio increases. Cement to aggregate ratio of 1:4 or lower provides high strength and durable blocks.

Replacing the washout blocks with good concrete blocks is recommended.

Fig. 6.10 Water curing in a block manufacturing yard

6.1.7 *Insufficient Strength of Interlocking Blocks*

A blend of coarse and fine aggregate is required to obtain the sufficient strength for interlocking blocks. A failed block shown in Fig. 6.11, has not been prepared using a sufficient amount of coarse aggregate in the mix design.

Fig. 6.11 Failed blocks

6.1.8 Lack of Edge Restraints

Edge restraints are important elements in a CBP road, to maintain the interlocking action. It has been found that formation of hinges and transferring of load laterally will be effective with proper edge restraints. Figure 6.12 shows a section of the road without an edge restraint (edge restraint has not been placed after the reparation of a utility). It can be seen in Fig. 6.13 that a concrete block near the edge restraint has displaced and broken due to loading. Washing out of the sand bed can also be observed (Fig. 6.14).

Figure 6.15 shows a failed curb line (Edge restraint) of a road. Edge restraints should be placed with a proper foundation, lateral support and behind the joints of precast curbs for its stability, and to resist the lateral load transfer through concrete haunching interlocking mechanism.

6.1.9 Poor Support from the Base Layer

Proper base/subbase layer thicknesses should be designed, depending on the subgrade condition. Excessive deformation is a result of the base/subbase failure. It is shown in Fig. 6.16.

Fig. 6.12 No edge restrain in the section of the road

Fig. 6.13 Wider gap and failed blocks

Fig. 6.14 Washout of the sand bed

Fig. 6.15 Poor support from edge restrain

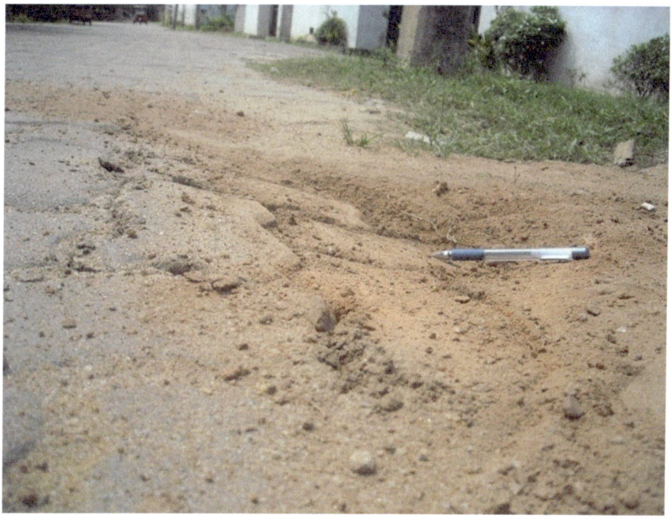

Fig. 6.16 Base failure

6.1.10 Faulting of Blocks

The level of difference allowed between two individual blocks is 2 mm for the sake of better appearance, workability and the facilitation of the surface runoff. There are several reasons for faulting, in a CBP road. They are; (1) poor compaction of the sand bed, (2) non uniform sand bed, (3) poor construction practices and (4) lack of quality control measures in place. Figure 6.17 shows a road section with more than 2 mm of faulting.

6.1.11 Rigid Joints (Mixing Concrete on Road Surface)

Mixing of concrete on block surfacing has detrimental effects on the interlocking mechanism, and transfers the hinges in CBP to rigid joints. Figure 6.18 shows a location on a CBP road in which concrete was mixed. As a result of this, blocks can fail, as shown in Fig. 6.19.

Fig. 6.17 Faulting of a CBP road

Fig. 6.18 CBP section with rigid joints (filled with cement mortar)

Fig. 6.19 Failure of rigidly set brick pavers at Bellevue Metro, Washington State, US ('John Knapton's Pictorial Guide to Paving through the Ages.', n.d.)

6.2 Maintenance Procedure

6.2.1 Removal of the Paving Blocks

Take off the blocks as shown in the Figs. 6.20, 6.21, 6.22 and 6.23. Usually, it is necessary to break the first few blocks for the easy removal of the remainder.

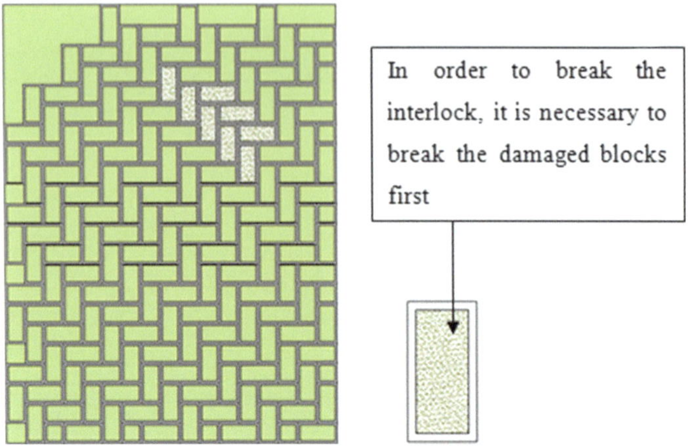

Fig. 6.20 Remove the blocks in the indicated order

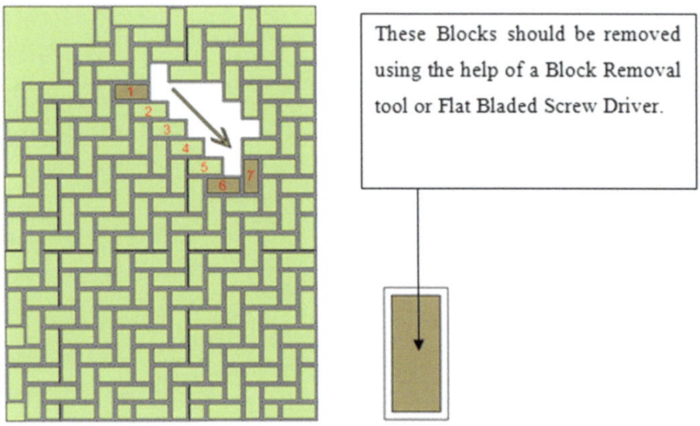

Fig. 6.21 Remove the blocks as in the indicated order

Clean the blocks to remove the sand and other detritus materials before stacking or re-using the blocks.

Stack the damaged and undamaged blocks separately. The damaged blocks can be reused as cut blocks.

Dispose the block laying material which cannot be reused.

6.3 Repair or Reinstate the Lower Pavement Layers

After the required area has been opened up, carry out the necessary reinstatement to the base layers. The required base thickness has to be achieved to get the finished

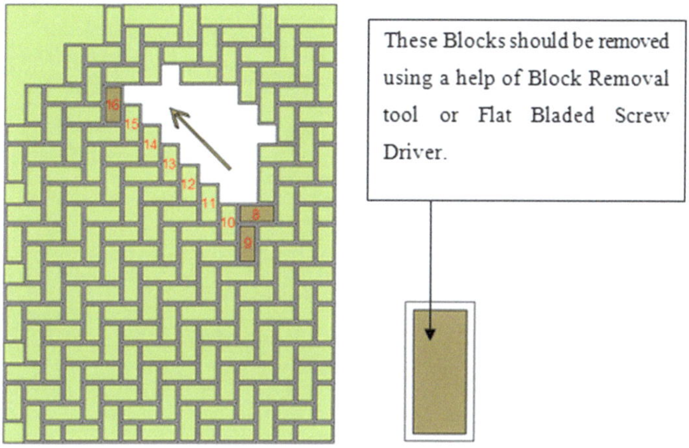

Fig. 6.22 Remove the blocks as in the order shown

level of the block layer. Additional care should be taken to select the material, and compaction and construction methods should be followed correctly, to avoid the future settlement. Ensure that the block paving layer is fully supported by the sand bed after opening up the lower layers as shown in Figs. 6.24 and 6.25.

Place and spread the correct laying course materials. Compact the new laying course material with the mechanical rammer until it meets the requirements.

Place the temporary screed rails at a maximum of 3 m centers to form a screeding datum. Check the rail levels by placing a block on top of each rail, and pulling a string line from each side of the existing pavement, across the top of the block. Ensure that the block is approximately 5 mm above the string line. Pack laying course material into any voids under the temporary screed rail to ensure that it does not deflect during screeding. Remove temporary screed rails, replace with laying course material and hand screed.

6.4 Select Replacement Blocks

Check whether the replacement blocks require the size compatibility over the existing blocks. It can be checked by laying 20 existing, and 20 new replacement blocks side by side, ensuring that they are packed tight together, as shown in Fig. 6.26. If the cumulative length of the replacement blocks is same or shorter, it would indicate that these blocks are smaller, and should fit easily into the reinstated area. If the cumulative length of the replacement blocks is longer, it would indicate that these blocks are larger, and that may cause difficulties when fitting these blocks into the reinstatement.

If such difficulties are encountered, a thin slice from some blocks shall be saw-cut to make them fit.

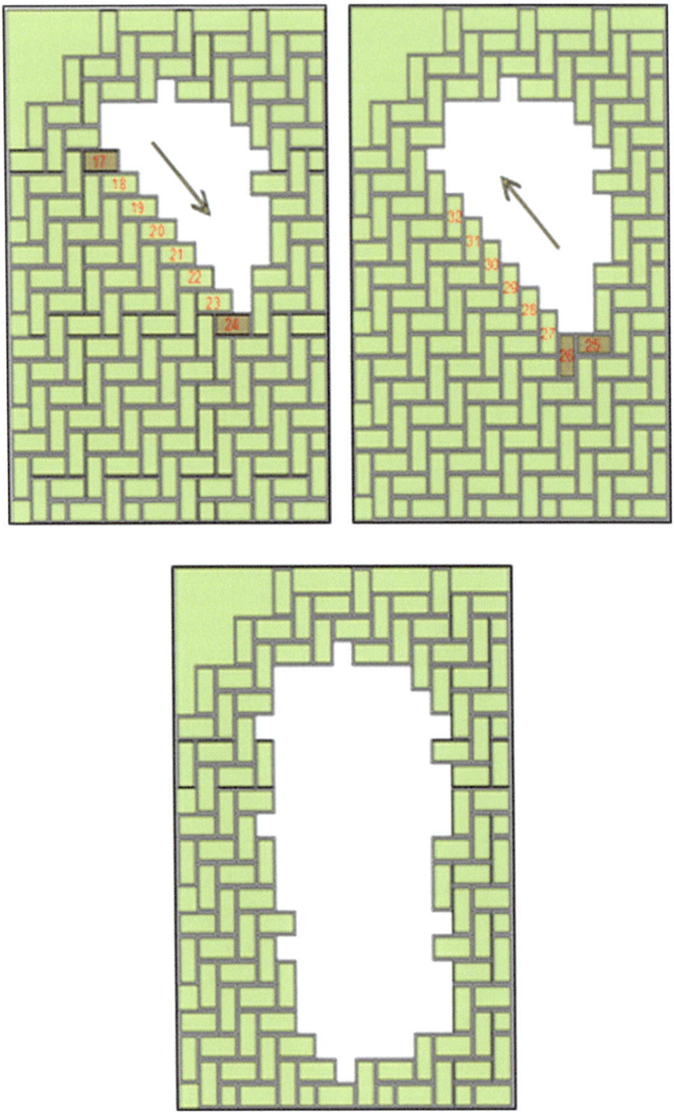

Fig. 6.23 Continue until the required area has been opened up

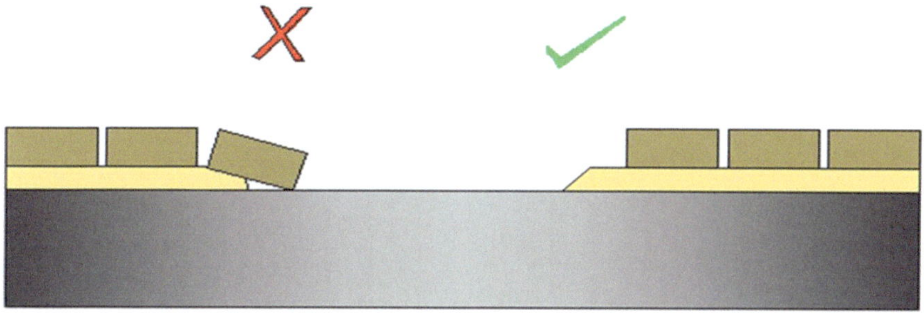

Fig. 6.24 Temporary screed rails

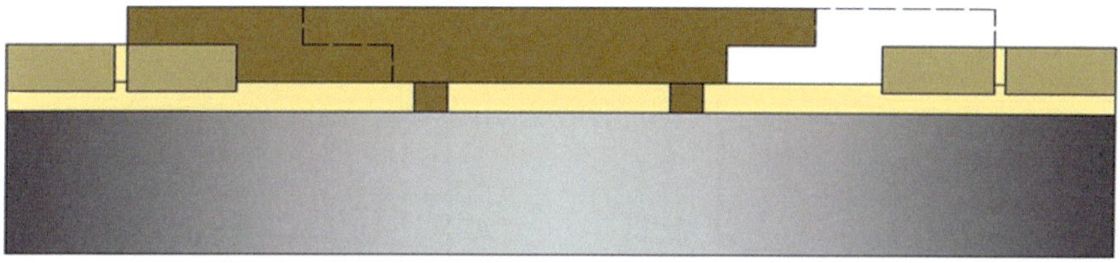

Fig. 6.25 Block paving layer fully supported by the sand layer

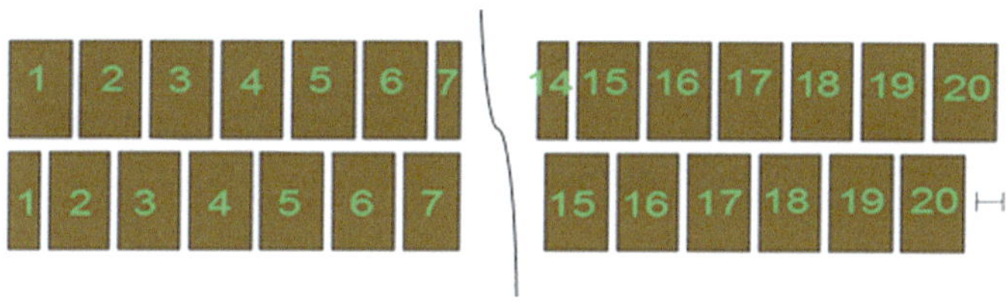

Fig. 6.26 Replacement of blocks

6.5 Laying of Blocks

6.5.1 Selection of an Effective Laying Angle

It was found that blocks should be laid in 90° or 0° for better performance. Laying of concrete blocks in an acute angle tend to load block edges and corners, leading to the differential settlement of the blocks as shown in Fig. 6.27. Figure 6.27a, b are an adjacent section of a CBP road located in Kirulapane, Sri Lanka. Relaying of blocks with 90° or 0° is recommended.

Fig. 6.27 Effect of poor laying angle

6.5.2 *Procedure of Laying Blocks*

Check whether the existing blocks around the perimeter of the reinstatement have moved or not. If necessary adjust or remove. Clean the surface of the pavement area to be re-laid. Avoid detritus from the joints, as this would affect the alignment or joint spaces.

Place the blocks in the required pattern as shown in Figs. 6.28, 6.29 and 6.30, with the joint space without disturbing the sand bed.

Continuously monitor the alignment of the pattern and check the alignment to the existing pavement in two directions at 90° to each other.

Adjust the alignment of the blocks to ensure that they fit together, with the aid of the alignment tools.

Compact the laying course by a suitable plate compactor over the laid blocks at least twice. Ensure that around the perimeter of the reinstatement, the compactor traverses half on the existing and half on the newly laid blocks.

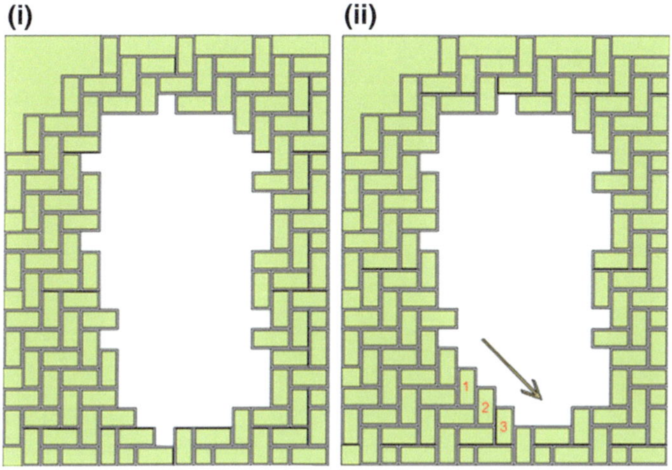

Fig. 6.28 Laying of blocks (i) and (ii)

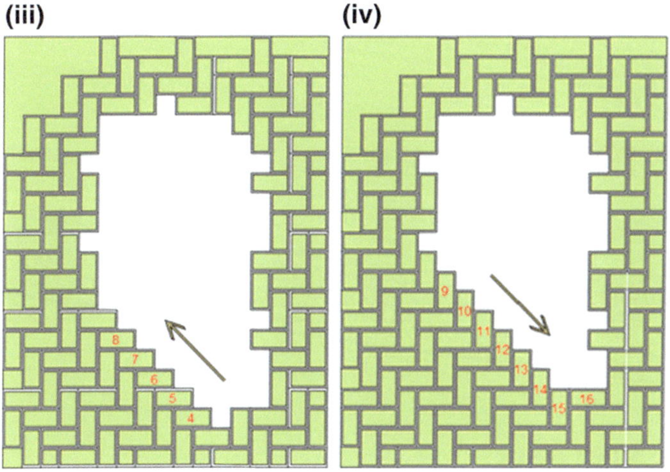

Fig. 6.29 Laying of blocks (iii) and (iv)

6.6 Check Joint Width and Block Heights

Check whether the Block alignment and joint widths are within the correct range.

Check whether the height of the Block layer has been obtruded, as shown in Fig. 6.31, and adjust if necessary by correcting the level of the laying course. If the laying course has been disturbed, re-compact after replacing the blocks (Fig. 6.31).

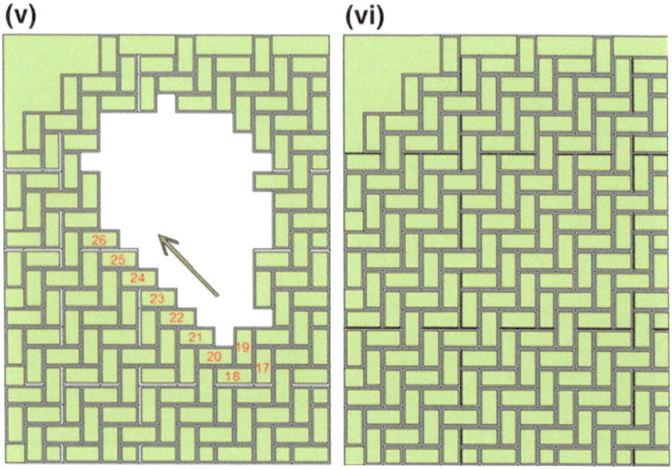

Fig. 6.30 Laying of blocks (v) and (vi)

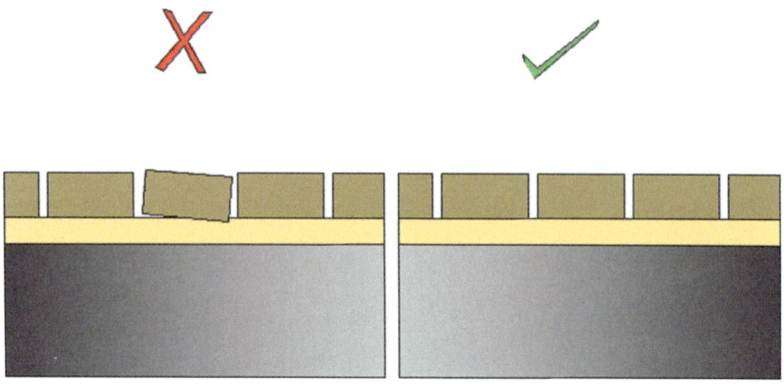

Fig. 6.31 Checking the block height

6.7 Fill Block Joint

Fill the block joints only after the alignment, joint widths and block heights are acceptable.

Select the correct joint filling material and compact by running the plate compactor over the laid blocks, ensuring the compactor traverses half on the existing and half on the newly laid blocks around the perimeter of the reinstatement.

At the end of each day, all the joints shall be filled with joint fill material.

6.8 Final Check

Check whether the alignment, joint width and block height are all in an acceptable condition and correct if necessary. Remove the debris and clean the area.

Appendix A
Equivalent Design Method

W. Mampearachchi, *Handbook on Concrete Block Paving*,
https://doi.org/10.1007/978-981-13-8417-2

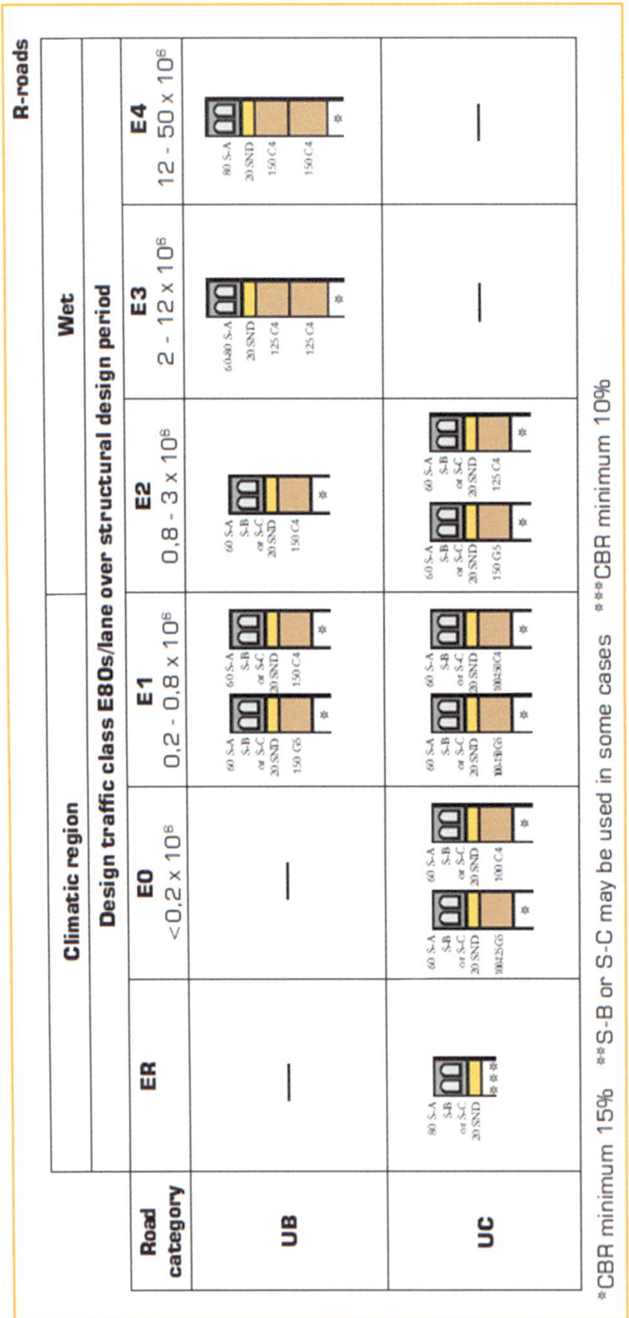

Concrete Manufacturing Association (2004)

Appendix B
Subgrade Improvement for Medium Traffic Level

Subgrade category	Medium traffic		
	TM3	TM2	TM1
S1	A80 + 200 GC[a] A80 + 180 SC + 200 ZN A80 + 180 MBB A80 + 100 MBB + 250 ZN A80 + 140 HM	A80 + 220 GC A80 + 200 SC + 250 ZN A80 + 210 MBB A80 + 140 MBB + 250 ZN A80 + 110 MBB + 150 SC[1] A80 + 160 HM A80 + 100 HM + 250 ZN	A80 + 200 GC + 250 ZN A80 + 260 MBB A80 + 190 MBB + 250 ZN A80 + 130 MBB + 200 SC A80 + 140 HM + 250 ZN A80 + 140 HM + 200 SC
S2	A80 + 110 MBB A80 + 160 SC	A80 + 200 SC A80[a] + 140 MBB A80 + 80 MBB + 200 ZN A80 + 110 HM	A80 + 160 SC + 200 ZN A80 + 180 GC A80 + 180 MBB A80 + 110 MBB + 250 ZN A80 + 140 HM A80 + 100 HM + 150 ZN
S3	A80 + 150 ZN	A80 + 80 MBB	A80 + 110 MBB A80 + 160 SC

[a]A80 + 200 GC—CBP of 80 mm thick + 30 mm sand and a base of 200 mm GC

W. Mampearachchi, *Handbook on Concrete Block Paving*,
https://doi.org/10.1007/978-981-13-8417-2

Appendix C
Design Specifications

Design specification is related to low volume roads and the size of the block shall be 225 mm × 100 mm × 75 mm.

CNSA (msa)	0.1–0.3	
Subgrade CBR (%)	Capping layer (mm) CBR 8+	Soil sub-base (SSB) Type 1 (mm) CBR 20+
2	300	250
3–4	200	200
5–7	–	200
8–14	–	125
15–19	–	100
20	–	–

Research and development division Road Development Authority July 2006

W. Mampearachchi, *Handbook on Concrete Block Paving*,
https://doi.org/10.1007/978-981-13-8417-2

Appendix D
Edge Restraints

W. Mampearachchi, *Handbook on Concrete Block Paving*,
https://doi.org/10.1007/978-981-13-8417-2

Edge restraints for low volume roads

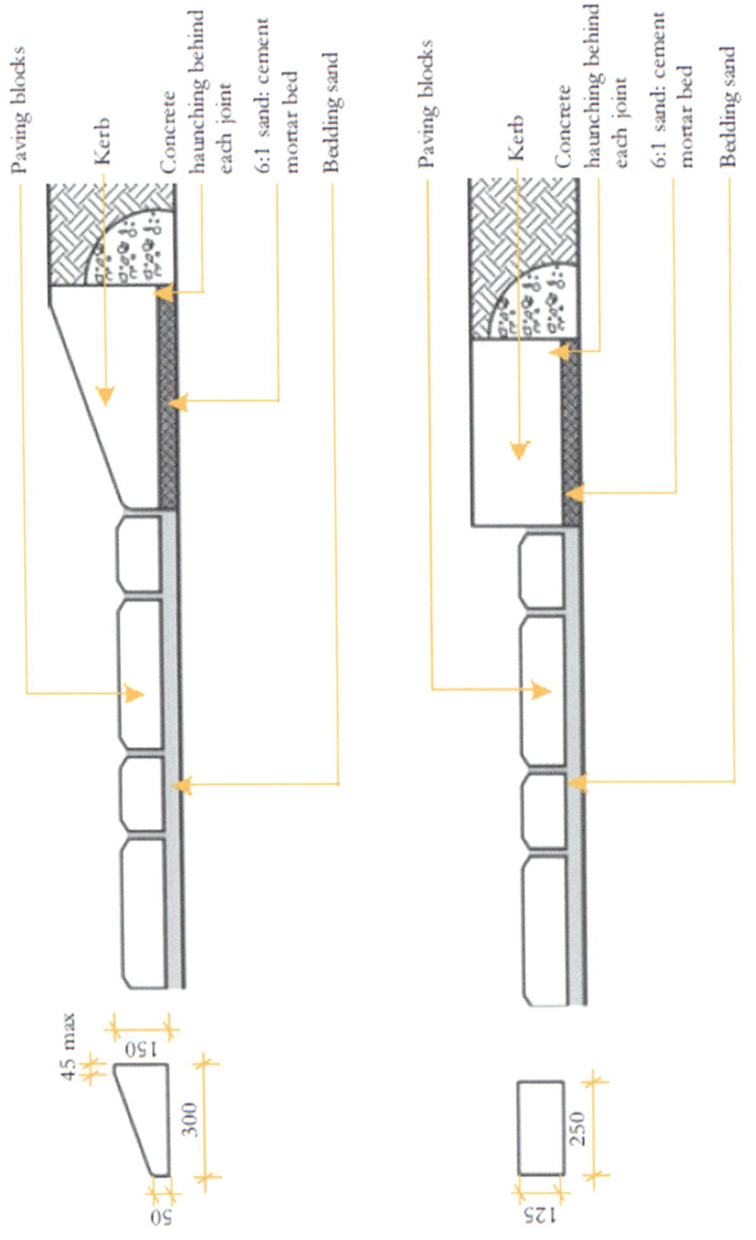

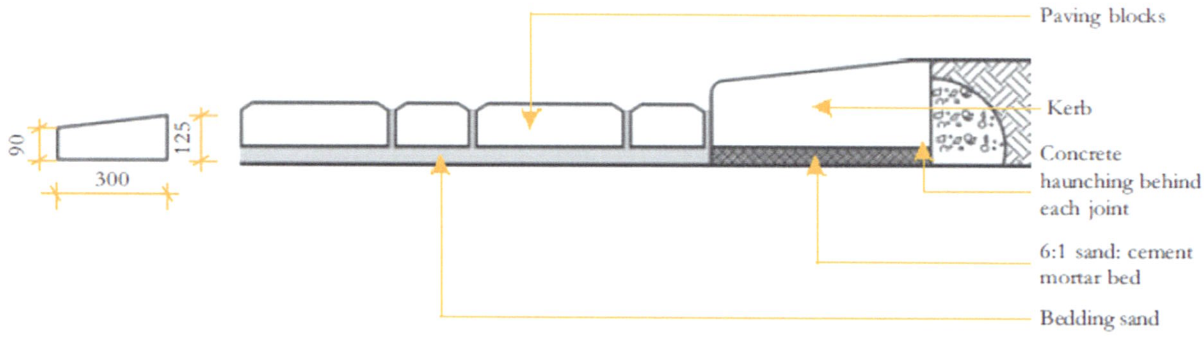

Paver thickness 60 mm
Precast concrete kerbs should comply with SANS 927-2003
Kerbs are generally supplied in lengths of 0.33 m or 1.0 m

Concrete manufacturing association (2004)

Edge restraints for Heavy Traffic roads

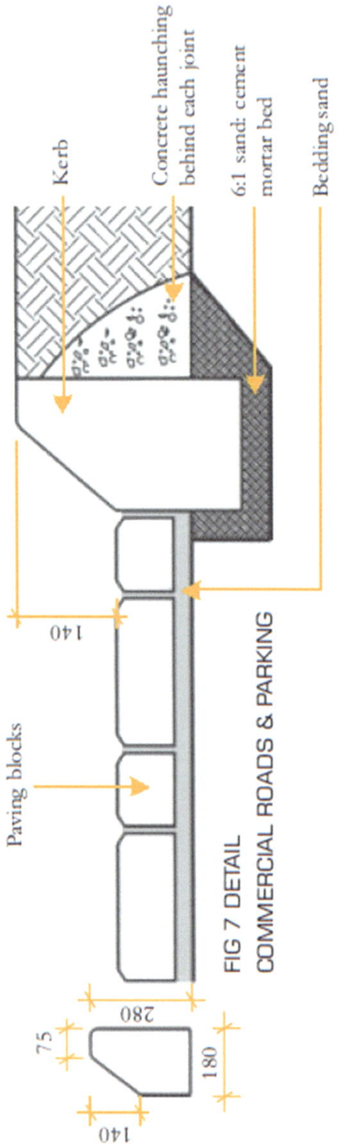

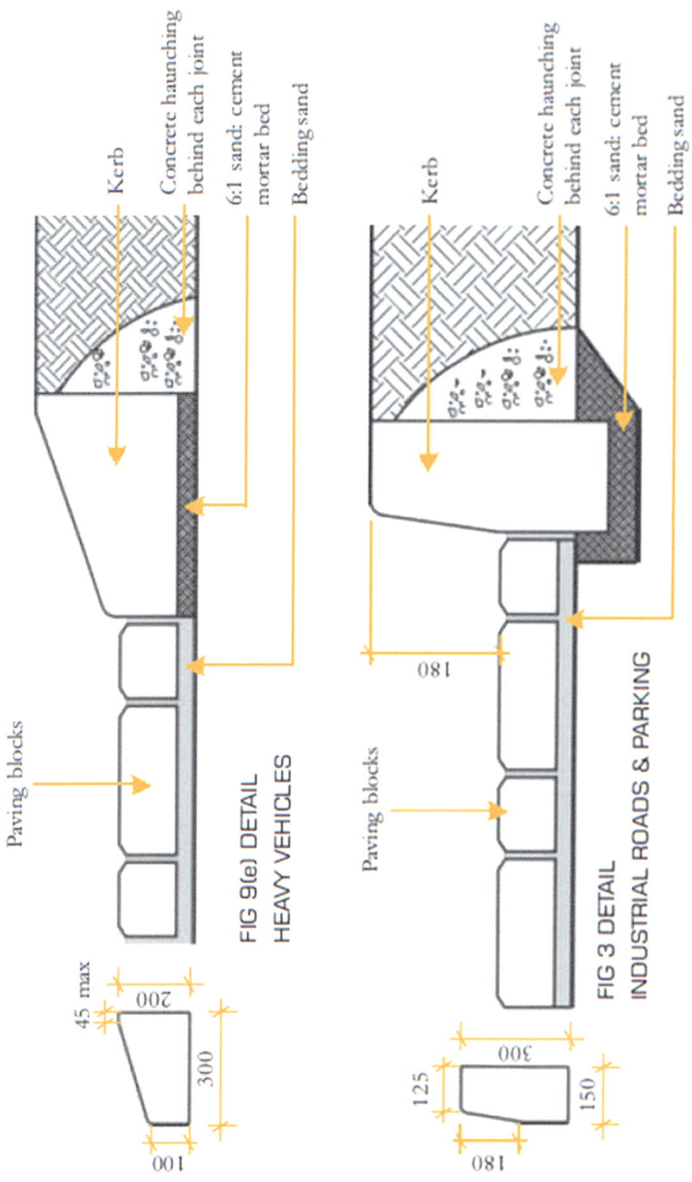
Kerb
Concrete haunching behind each joint
6:1 sand: cement mortar bed
Bedding sand
Paving blocks
FIG 9(e) DETAIL
HEAVY VEHICLES
45 max
200
300
100
Kerb
Concrete haunching behind each joint
6:1 sand: cement mortar bed
Bedding sand
180
Paving blocks
FIG 3 DETAIL
INDUSTRIAL ROADS & PARKING
300
125
150
180

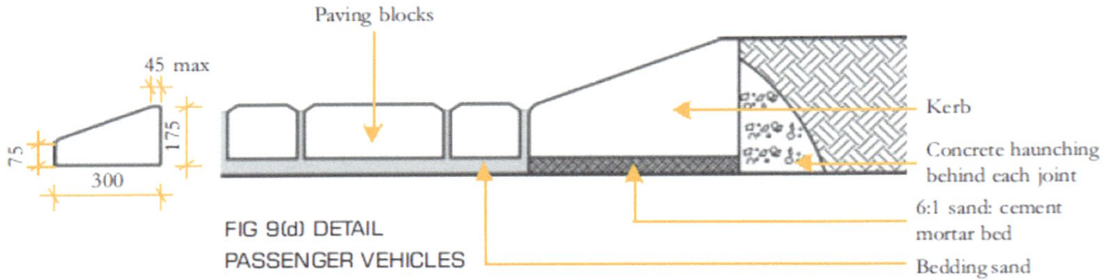

Pavers thickness 80 mm
Precast concrete kerbs should comply with SANS 927-2003
Kerbs are generally supplied in lengths of 0.33 m or 1.0 m
For extra heavy duty areas, continuous concrete hunching should be placed behind kerb/edge restraint

Concrete manufacturing association (2004)

References

Abril, J. M., Irastotza, L., & Josa, A. (1994). *A manual for concrete block pavement design* (pp. 141–148). Presented at the Concrete Block Pavement Second International Workshop, Oslo, Norway.

ASTM International. (2018). *Standard specification for solid concrete interlocking paving units* (No. ASTM C936 / C936M).

British Standards Institution. (n.d.). *1993 Precast, unreinforced concrete paving blocks—requirements and test methods* (No. BS 6717-1). London.

Clark, A. J. (1978). Block paving-research and development. *Concrete, 12*(7).

Concrete Manufacturing Association. (2004). *Concrete block paving book* (Vol. 1–4). Pretoria, South Africa: Concrete Manufacturing Association.

Cook, M. D., Ley, M. T., & Ghaeezadah, A. (2014). A workability test for slip formed concrete pavements. *Construction and Building Materials, 68,* 376–383.

Ghafoori, N., & Mathis, R. P. (1997). Scaling resistance of concrete paving block surfaces exposed to deicing chemicals. *Materials Journal, 94*(1), 32–38.

Ghafoori, N., & Mathis, R. (1998). Prediction of freezing and thawing durability of concrete paving blocks. *Journal of Materials in Civil Engineering, 10*(1), 45–51.

Hettiarachchi, H. A. C. K., & Mampearachchi, W. K. (2018). Optimization of *Aggregates for sustainable low cement concrete for interlocking concrete block pavers* (pp. 2894–2900). Presented at the World Sustainable Built Environment Conference (WSBE17), Hong Kong.

Hettiarachchi, H. A. C. K., & Mampearachchi, W. K. (2019). Validity of aggregate packing models in mixture design of interlocking concrete block pavers (ICBP). *Road Materials and Pavement Design, 20*(2), 462–474.

Institution, I.S. (2006). *Precast concrete blocks for paving* (No. IS:15658:2006).

Institution, S.L.S. (2011). *Sri Lanka standard specification for concrete paving blocks* (No. SLS 1425 Part 2:2011).

John Knapton's Pictorial Guide to Paving through the Ages. (n.d.). Retrieved 3 April 2019, from http://www.john-knapton.com/paving.htm.

Knapton, J. (1976). *The design of concrete block roads* (No. 42.515). Wexham Springs, UK: Cement and Concrete Association.

Knapton, J., & O'Grady, M. (1983). Structural behaviour of concrete block paving. *Journal Concrete Society.*

Kumbhar, P. D., & Murnal, P. B. (2012). Assessment of suitability of existing mix design methods of normal concrete for designing high performance concrete mixes. *International Journal of Civil and Structural Engineering, 3*(1), 158–167.

W. Mampearachchi, *Handbook on Concrete Block Paving*,
https://doi.org/10.1007/978-981-13-8417-2

Mampearachchi, W. K., & Gunarathna, W. P. H. (2010). Finite-element model approach to determine support conditions and effective layout for concrete block paving. *Journal of Materials in Civil Engineering, 22*(11), 1139–1147.

Mampearachchi, W. K., & Senadeera, A. (2014). Determination of the most effective cement concrete block laying pattern and shape for road pavement based on field performance. *Journal of Materials in Civil Engineering, 26*(2), 226–232.

Miura, Y., Takaura, M., & Tsuda, T. (1984). Structural design of concrete block pavements by CBR method and its evaluation. In *Proceedings of 2nd International Conference on Concrete Block Paving* (pp. 152–157). Delft, The Netherlands: Delft University of Technology.

Murat Algin, H. (2007). Interlock mechanism of concrete block pavements. *Journal of Transportation Engineering, 133*(5), 318–326.

Panda, B. C., & Ghosh, A. K. (2001). Source of jointing sand for concrete block pavement. *Journal of Materials in Civil Engineering, 13*(3), 235–237.

Panda, B. C., & Ghosh, A. K. (2002a). Structural behavior of concrete block paving. I: sand in bed and joints. *Journal of Transportation Engineering*, *128*(2), 123–129.

Panda, B. C., & Ghosh, A. K. (2002b). Structural behavior of concrete block paving. II: concrete blocks. *Journal of Transportation Engineering*, *128*(2), 130–135.

SABS. (2006). *Concrete paving blocks* (No. S.A.B.o.S. SANS 1058:2006).

SABS. (2012). *Concrete paving blocks* (No. S.A.B.o.S. SANS 1058:2012).

Shackel, B. (1980). The performance of interlocking block pavements under accelerated trafficking. In *Proceedings of 1st International Conference on Concrete Block Paving* (pp. 113–120). Newcastle-upon-Tyne, UK.

Shackel, B., O'Keeffe, W., & O'Keeffe, L. (1993). Concrete block paving tested as articulated slabs. In *Fifth International Conference on Concrete Pavement Design and Rehabilitation Purdue University, School of Civil Engineering; Federal Highway Administration; Portland Cement Association; Transportation Research Board; Indiana Department of Transportation; Federal Aviation Administration; and American Concrete Pavement Association* (Vol. 1, pp. 89–95).

Shahriar, A., Amin, A. F. M. S., Ahmad, S., Hossain, M. Z. (2005). *Some Thoughts on ACI Method of Concrete Mix Design, ICRAMME 05, Proceedings of the International Conference on Recent Advances in Mechanical & Materials Engineering*, 30–31 May 2005, Kuala Lumpur, Malaysia, Paper No. 162. [ISBN 983-41728-2-6].

Taylor, P., Bektas, F., Yurdakul, E., & Ceylan, H. (2012). Optimizing cementious content in concrete mixtures for required performance. Ames, IA, USA: National Pavement Technology Center.

Tsohos, G. H., & Iliou, N. E. (1994). *Design of concrete block pavements*. Presented at the Concrete Block Pavement Second International Workshop, Oslo, Norway.

MIX
Papier aus verantwortungsvollen Quellen
Paper from responsible sources
FSC® C105338

If you have any concerns about our products,
you can contact us on
ProductSafety@springernature.com

In case Publisher is established outside the EU,
the EU authorized representative is:
Springer Nature Customer Service Center GmbH
Europaplatz 3, 69115 Heidelberg, Germany

Printed by Libri Plureos GmbH
in Hamburg, Germany